# THIS
# CHILD
# WILL BE
# GREAT

# THIS
# CHILD
## WILL BE
# GREAT

*Memoir of a Remarkable Life by*
*Africa's First Woman President*

## ELLEN JOHNSON SIRLEAF

HARPER
*An Imprint of* HarperCollins*Publishers*
www.harpercollins.com

THIS CHILD WILL BE GREAT. Copyright © 2009 by Ellen Johnson Sirleaf. All rights reserved. Printed in the United States of America. No part of this book may be used or reproduced in any manner whatsoever without written permission except in the case of brief quotations embodied in critical articles and reviews. For information, address HarperCollins Publishers, 10 East 53rd Street, New York, NY 10022.

HarperCollins books may be purchased for educational, business, or sales promotional use. For information, please write: Special Markets Department, HarperCollins Publishers, 10 East 53rd Street, New York, NY 10022.

All photographs courtesy of the Johnson Sirleaf family.

FIRST EDITION

Designed by Eric Butler

Library of Congress Cataloging-in-Publication Data is available upon request.

ISBN: 978-0-06-135347-5

09  10  11  12  13   OV/RRD   10 9 8 7 6 5 4 3 2 1

*To all the people of Liberia*
*who have suffered so much and now*
*look forward to reclaiming the future.*

*And in memory of my mother,*
*Martha Cecelia Johnson,*
*who instilled in us the value of*
*hard work, honesty, and humility.*

# CONTENTS

CONTENTS

THIS

CHILD

WILL BE

GREAT

# PROLOGUE

---

IF ASKED to describe my homeland in a sentence, I might say something like this: Liberia is a wonderful, beautiful, mixed-up country struggling mightily to find itself.

Given more space, however, I would certainly elaborate.

Liberia is some 43,000 square miles of lush, well-watered land on the bulge of West Africa, a country slightly larger than the state of Ohio, a lilliputian nation with a giant history. It has a population of 3.5 million people from some sixteen ethnic groups speaking some sixteen indigenous languages plus English. It has never known hurricanes, tornadoes, earthquakes, droughts, or other natural disasters, only the occasional flood and the more frequent havoc wreaked by man. Liberia is complicated. To paraphrase Winston Churchill, Liberia is a conundrum wrapped in complexity and stuffed inside a paradox. Then again, it was born that way.

The first inhabitants of the region now known as Liberia may have been Jinna, or pygmies, according to the Liberian historian Abayomi Karnga. Soon came the Gola, whom many historians believe to be the first traveling settlers of the land. Gola legend has it that the tribe left Central Africa and moved toward the coast in search of land. Ruthless fighters, they went through and not around the tribes in their path, according to the book *Notes from the New Liberia: A Historical and Political Survey.* The Gola and the Kissi belong to the Mel (West Atlantic) ethnolinguistic group.

The Mande linguistic group, made up of the Mandingo, the Vai (one of only a few African tribes to have developed a script), the Gbandi, the Kpelle, the Loma (who also had a script), the Mende, the Gio, and the Mano peoples, is believed to have entered the area from the northern savannas in the fifteenth century. The third major group, the Kwa linguistic group, includes the Bassa, Dei (Dey), Grebo, Kru, Belle (Kuwaa), Krahn, and Gbee peoples, found mostly in the southern and eastern parts of Liberia.

All of these groups were living in the land when the final group of settlers began to arrive. These were the Americo-Liberians.

As early as the 1700s, the idea of sending New World slaves "back" to Africa rose in the hearts and minds of British abolitionists, who saw, in the establishment of a colony for former slaves, a means of ending the slave trade—and, eventually, slavery itself.

During the American Revolutionary War, African slaves in the American colonies were promised freedom if they sided with the British. Many did, fighting valiantly. When the war ended, several hundred of these fighters gathered their families and fled the country with the departing British troops. After some wandering, they were settled, with British backing, along the coast of West Africa in what is now the country of Sierra Leone. Although most of those first, early settlers perished of malaria and yellow fever, subsequent attempts at settlement took firmer hold. The first of these was established in 1792. The settlers called their new city Freetown.

Meanwhile, a similar conversation was taking place across the pond in the newly born United States of America. Paul Cuffee was the free-born son of a former slave father and a Native American mother. A prominent Quaker, Cuffee became a sailor and successful shipowner who opened the first integrated school in Massachusetts and later began advocating to settle freed slaves in Africa. In 1816, at his own expense but with the backing of the British government and some members of the U.S. Congress, Captain Cuffee took thirty-eight American blacks to Freetown and settled them there. He intended to make this voyage an annual affair, but his untimely death in 1817 put an end to the plans.

Still, Cuffee had reached a large audience with his procoloniza-tion arguments. Not all supporters of the idea, however, had arrived at their destination from the same starting point.

Some abolitionists took up the cry as part of their attack on slavery, while some religious adherents thought the idea an excellent means of spreading Christianity to "the dark continent." Some white Ameri-cans, considering themselves pragmatists, thought black people would simply be happier and fare better in Africa, where they could live free of the racial discrimination that gripped America. Others simply wanted to rid the new country of all black people as soon as possible.

In this last group stood the statesman Thomas Jefferson. Writing in his 1781 *Notes from the State of Virginia*, Jefferson advocated gradu-ally emancipating all slaves and shipping them to Africa, along with free blacks and possibly those of mixed heritage. Jefferson—an author of the American Declaration of Independence and its famous, stirring words "We hold these truths to be self-evident, that all men are created equal, that they are endowed by their Creator with certain unalienable Rights"—was, of course, a slave owner himself. He considered black people inferior to whites and warned that their continuing presence was a threat to the young nation he had helped to found. They could not simply be freed and allowed to remain in the United States, Jeffer-son warned. They had to go.

"Among the Romans emancipation required but one effort," he wrote. "The slave, when made free, might mix with, without staining the blood of his master. But with us a second is necessary, unknown to history. When freed, he is to be removed beyond the reach of mixture."

This, then, was the backdrop against which the American Colo-nization Society was founded in 1816 in Washington, D.C. Among the founders were many prominent early American leaders, including Daniel Webster, Francis Scott Key, Henry Clay, and Bushrod Wash-ington, an associate justice of the Supreme Court and the nephew of George Washington himself (and after whom Bushrod Island in Monrovia is named).

The society began raising funds to establish a colony in Africa.

Four years later, in January of 1820, the ship *Elizabeth* set sail from New York. On board were eighty-six free American blacks from New York, Philadelphia, Washington, and Virginia—more than half of them women and children—along with three agents of the ACS. It took six weeks to cross the Atlantic, and the eager immigrants landed first on Sherbo Island off the coast of present-day Sierra Leone, intending to use the spot as a stopping ground while they searched for permanent accommodations on the nearby mainland.

But disease and fever laid waste to much of the group; by May 1820, all three agents and nearly a quarter of the would-be settlers were dead. Those who survived fled to Freetown to recover and repair.

A year later the society sent a new agent, Dr. Eli Ayres, to explore the coast and negotiate with local Dey and Bassa chiefs for suitable land for a settlement. Under the direction of President James Monroe, Dr. Ayres enlisted the help of Lieutenant Robert Stockton, captain of the USS *Alligator*, a naval ship patrolling the West African coastline in cooperation with His Majesty's Navy's slave-trading blockade. Together Ayres and Stockton settled their sights on a slip of land two hundred or so miles south of Freetown known as Cape Montserado, or Mesurado. Agents of the ACS had previously tried to buy this very land, but the local Bassa chief, King Peter, had declined to sell.

This time, however, Stockton declined to take no for an answer. When King Peter again seemed reluctant to sell, Stockton persuaded him with a pistol to the head. Thus, in December 1821, the ACS gained a toehold in Africa in exchange for some $300 worth of goods, including muskets, gunpowder, nails, beads, tobacco, shoes, soap, and rum. Stockton also promised that the new immigrants would not interfere with the thriving local slave trade.

Soon afterward, the surviving settlers from the *Elizabeth*, replenished by a fresh ship full of immigrants, took possession of the land. They gave thanks, dubbing their new home Providence Island, and then moved immediately to secure the adjacent mainland. There the first permanent settlement, originally called Christopolis, was carved from the thick forest. In 1824, the settlement's name was changed to Monrovia, in honor of U.S. President (and ACS member) James Monroe, and the colony as a whole became the Commonwealth of Liberia.

But the colony, being neither a sovereign nation nor a bona fide colony of a sovereign government (the United States refused to formally designate it as such) faced ever-increasing political threats from foreign governments, especially Britain. So on July 26, 1847, the country declared independence in a document that referenced the oppression of African Americans in the United States and stated that such persons were being driven to make new lives for themselves in Africa.

And so Liberia was born.

Britain was one of the first governments to recognize the new nation; the United States did not recognize Liberia until the American Civil War.

Nonetheless, that initial settler group, and those who followed, came to call themselves Americo-Liberians. It was an appropriate name, for this band of hopeful immigrants was far more American than African and intended to remain that way. They had adopted the cultures, traditions, and habits of the land of their birth, and these they brought with them when they came.

The colonists spoke English and retained the dress, manners, housing, and religion of the American South. They created symbols for this new nation that reflected, to a sometimes startling degree, their American identity and emigrant sensibility. The Liberian flag mimics so closely the American flag that it is easy to mistake it for the latter; the only difference is the number of stars and stripes. The Liberian seal features not only a palm tree, representing our vast natural resources, but a sailing ship to represent the settlers who came from across the seas.

The Liberian motto is "The love of liberty brought us here." Several passages of our original Declaration of Independence might sound familiar to American ears: "We the People of the Commonwealth of Liberia, in Africa, acknowledging with devout gratitude, the goodness of God . . . do, in order to secure these blessings for ourselves and our posterity, and to establish justice, insure domestic peace, and promote the general welfare, hereby solemnly associate, and constitute ourselves

a Free, Sovereign and Independent State, by the name of the REPUB-
LIC of LIBERIA, and do ordain and establish this Constitution for
the government of the same."

The settlers of modern-day Liberia decided they would plant their
feet in Africa but keep their faces turned squarely toward the United
States. This stance would trigger a profound alienation between themselves and the indigenous peoples upon whose shores they had arrived
and among whom they would build their new home. Alienation would
lead to disunity; disunity would lead to a deeply cleaved society.

That cleavage would set the stage for all the terror and bloodshed
to come.

# CHAPTER 1

# THE BEGINNING

WHEN I was just a few days old, an old man came to visit my parents, to see the new baby and to offer his good wishes, as people did both then and now in my country and everywhere. My mother brought the old man into the room where I lay kicking and cooing on the bed. As the story goes, the old man took one look at me and turned to my mother with a strange expression on his face.

"Oh, Martha," he said. "This child shall be great. This child is going to lead."

My mother and sister and I used to laugh whenever my mother told this story. We would laugh and laugh and laugh, because at many of the junctures in which she recalled the words of the wise old man my life seemed anything but great. Perhaps I was watching all my friends go off to college abroad while I stayed at home in Monrovia, trapped with an abusive husband, four young sons, and no future in sight. Perhaps I was struggling to pursue my education, build my career, and divorce that husband without losing everything I had. Or perhaps I was being hauled off to prison by order of my nation's president—or maybe even plotting an escape into exile to save my life.

"Where's all this greatness that was predicted?" my mother would ask. Sometimes she laughed, sometimes she cried. Always she prayed. "Where's that old man now?"

Over the years and as the path of greatness unfolded, whenever I reflected on the prophecy of the old man, my scientific orientation of self-determination would clash with the Presbyterian teachings of predestination I had received.

Which one, I have long wondered, is the way life really is?

EARLY ON during my historic 2005 campaign for the presidency of Liberia, rumors began to circulate about my ethnicity. My detractors began whispering that I was an Americo-Liberian, a descendant of one of those first American-born founders of our land—and thus a member of the elite class that had ruled our nation for long.

This was an explosive charge. Given the historic cleavage in our society and the long-standing divide between the elite settler and indigenous populations, many Liberians wanted nothing to do with another Americo-Liberian president. And although I was well known in my country—so well known that most people, including the swarms of children who would come out to greet me as I campaigned, simply called me "Ellen"—still, there was danger that the rumor would find traction. It could not be brushed off or ignored, not if I wanted to win. It was crucial that the people of Liberia know my background was not unlike their own. They needed to know where I was coming from.

In truth, my family exemplifies the economic and social divide that has torn our nation. But, unlike many privileged Liberians, I can claim no American lineage.

My paternal grandfather was a Gola chief of great renown. His name was Jahmale, sometimes called Jahmale the Peacemaker, and he lived, along with his eight wives, in the village of Julejuah, in Bomi County. Jahmale used to travel from his home village to the ocean, a distance of some twelve or fifteen miles that, in those days, took months and months of slow walking through the dense forests of coastal Liberia. During his travels he learned to speak the languages and dialects of the many peoples whose path he crossed and so became a kind of negotiator when troubles erupted between the indigenous people and the settlers in Monrovia.

In this way his reputation grew, and it was because of this renown that my grandfather was sometimes visited by Hilary Wright Johnson,

Liberia's eleventh president. Johnson was the first president of Liberia to be born in our country. He was also the son of Elijah Johnson, one of the original settlers.

At that time there were few roads in Liberia and none at all outside the capital. So when the president traveled into the hinterland to visit villages, he, along with his entourage, would be carried about in hammocks, welcomed with food and dance and celebration and perhaps the gift of a young woman as a wife. The president in turn brought excitement, gifts, and connection to the country's power base back in Monrovia. It was President Johnson who encouraged Jahmale to send my father to the city as a ward.

As with many aspects of Liberian society, the ward system, its history and legacy, is not simple to parse. Its origins seem to lie in a complex combination of tradition, expediency, and need; the motivations of its participants varied greatly, as did the way in which it was executed.

In the simplest explanation, the ward system flourished in early Liberia because it met the settlers' crucial need for cheap labor. Those early transplanted families, not having enough children themselves, needed help with the heavy housework of the nineteenth century: hauling water, collecting firewood and coal, cooking, cleaning, and tending crops.

At the same time, it was, in many villages, an African tradition for chiefs and wealthier villagers to have guardianship of children whose parents were either dead or too poor to care for them. The extended family system in Africa assumes that everyone is his brother's keeper; it is one of our strengths. Likewise, it was common at the time for chiefs who formed alliances with other tribes or chiefs to offer women as wives and children as wards to validate the agreements.

The American Colonization Society, recognizing how the tradition could be used to spread Christianity among the indigenous population, encouraged the settlers to take local children into their homes. In many cases these young people, once accepted into the family, were treated equally and given the same duties, responsibilities, and opportunities as the family's own biological offspring. Often settlers grew so fond of their wards that they provided for them in their wills, as did Samuel C. Coker, a settler farmer from Bensonville, who gave generous grants of lands to three of his wards—provided, he wrote, they remain "among the civilized elements."

But to say that every ward was warmly embraced by his or her new family would be untrue. In some cases wards were viewed mostly as a source of cheap labor, unpaid servants who were yours to treat or mistreat as you saw fit—slaves, essentially. This is the painful and unvarnished truth.

However, it must also be said that the majority of families, regardless of how discriminating or unjust, gave the wards in their care some opportunity for education. Limited, perhaps, but more than they would have had otherwise.

Even some of the best-treated wards suffered the indignity of having their names changed to better suit their new, nonindigenous status. My father, who was taken in by a family named McGrity, was given the last name of Johnson, after the president. His first name, Karnley, was westernized to Carney. At the age of fifteen or sixteen, Carney Johnson was born. In the name of assimilation and to be accepted, he had to experience a rebirth.

When I was a girl, my father would sometimes tell us stories of the years he had spent as a ward, performing chores, cleaning, collecting water and firewood. For the most part the McGritys were good to him and treated him well. But the line between him and the family was always clear, and he was punished if he was disrespectful or did something wrong. Once, for some infraction, he was stuffed into a large bag and hung over an open fire as punishment. He wasn't held there long enough to be injured, just long enough to make the point.

Yet my father benefited tremendously from his time as a ward. First and foremost, the McGritys sent him to school. He was given a solid primary education, something that would not have happened back in the village. With that education, the world opened to him.

In those days there were no law schools to speak of in Liberia; one became a lawyer either by going abroad or by apprenticing with a practicing attorney. My father did the latter. He became what was called a poor man's lawyer, a very successful one. He married and, like many Liberians, began looking toward a life in politics as a way to both serve his country and build his career. He was climbing the ladder of success in the Liberian legislature when, one day, walking through the streets of Monrovia, he spotted my mother and fell in love.

My MOTHER's story is similar to my father's, though she was much younger than he. Martha's mother, Juah Sarwee, was a native farmer and market woman from Greenville, Sinoe County, who fell in love with a trader from Germany. During the early part of the twentieth century, Germany was one of Liberia's major trading partners, and the country was full of German expatriates organizing the exports of coffee, palm oil, palm kernels, and piassava. But at the start of World War I, Liberia, eager to display its loyalty to the United States, declared war on Germany and expelled all German citizens. My maternal grandfather left Liberia, left his wife, left his young daughter. He was never heard from again. About her father my mother rarely spoke. It was a part of her life she never knew and never cared to find out about. She had other relatives in her village who also had German blood in their veins, and sometimes, at family gatherings, they would sit around trying to reminisce about a time that had passed. My mother would never join them. To her, what was gone was gone.

Only once in my life did I hear her mention her father with any sense of regret. It was many years after the wars, when Liberia and Germany had reestablished friendly relations and Germans were once again visiting our land. "I wish I had known my father," my mother said one day, reading some item in the newspaper. But it was just a passing comment. After making it, she folded the newspaper away and got on with her work.

Because of her father, my mother was a fair-skinned child with long, wavy hair. She could almost pass for white, and naturally she stood out in her village. She became an attraction of sorts, this child who was quite different from all the other children. Many settler families offered to take her into their homes, and eventually her mother, who was poor, illiterate, and struggling to feed them both, agreed.

The first family treated my mother very poorly, more as an indentured servant than as a foster child. She had no room of her own, not even a bed, but often slept on the table or even sometimes beneath it with the family animals. Her days were full of chores, and she was neither properly fed nor groomed. When word got out in Monrovia that this fair-skinned child was being mistreated by her settler family, a woman named Cecilia Dunbar, the wife of Charles Dunbar, stepped up and offered a settlement.

"Give her to me," she said. "I will take care of her."

The Dunbars were a very old and very prominent Liberian family, dating back to one of the first groups of settlers. One of the ancestors of Charles Dunbar was Charles B. Dunbar, one of three men who briefly led the nation as part of an executive committee during a presidential crisis in 1871.

Cecilia Dunbar had no children of her own, and no doubt my mother's fair skin was the source of at least part of her interest in the child. Nonetheless, it was a humanitarian gesture. My mother became, in effect, the only child of a well-off family, and needless to say it changed her life. She took the Dunbar name, went to school, and received the best education possible in Liberia. She was even sent abroad to further her studies for a year.

But she was still just a teenager the day my father passed the Dunbar house and spotted her in the yard. The sight of her stopped him cold.

"Oh," he said, struck by her hair, her figure, her overall loveliness. "Oh. I like you."

My mother, however, was not impressed at first. She turned her back to him, rushing into the house to tell Mrs. Dunbar that some older man was outside bothering her. By the time they got back outside, he was gone.

But he kept an eye on her, visiting sometimes and watching as she grew. It was four or five years later that he divorced his wife and visited Grandmother Cecilia to ask for my mother's hand in marriage. By this time my mother had softened toward him. He was a handsome man, my father—tall, brown-skinned, and stylish, with a special, jaunty way of walking that proclaimed his confidence. When he proposed to my mother and asked Grandmother Cecilia for her daughter's hand, both women were won over. Both women immediately said yes.

And so my parents, a son of a Gola chief from Bomi County and a daughter of a market woman from Sinoe, were married. I can imagine them as a young couple in Monrovia: hardworking, ambitious, eager to create a better life for themselves and their family. They would work hard and together create a better life.

And so they did, until a sad and sudden turn of events knocked our family off the ladder of success.

THE MONROVIA of today is a grand but wounded city, the bruised and battered capital of a bruised and battered land. The Monrovia of my youth was a different place: simpler in feel, smaller in scale. We loved it dearly, but in truth it was not much of a city, more like a large village by the sea. There were no public transportation system, telephones, or streetlights, and very few cars. We walked to school, walked to church, walked to our neighbors' houses. When we went out of town, we traveled in hammocks or canoes. Nor was the city dotted with the grand, imposing buildings later built by Presidents William Tubman, William Tolbert, and Samuel Doe and badly damaged during the wars. The Monrovia of my childhood was zinc houses and hilly dirt streets, papaya trees and cassava plants, flowering gardens and wooden outhouses: simple, friendly, close-knit. Home.

Monrovia was also, from the beginning, the undisputed seat of power of Liberia. In the years immediately following the founding of Monrovia, slave-owning states, hoping to rid themselves of the worrisome free blacks in their midst, formed their own societies and founded their own colonies, working independently of the ACS. In 1834, the Maryland State Colonization Society established its colony at Cape Palmas, some 250 miles south of Monrovia. Societies in Pennsylvania and New York worked together to settle the Edina and Port Cresson (later Bassa Cove) colonies in 1832, while groups in Virginia and Mississippi worked to establish their own colonies.

But it was the ACS that continued to flood the burgeoning country with settlers. Year after year they arrived, riding the brigs *Nautilus, Strong, Hunter,* and *Vine;* the ships *Cyrus, Norfolk,* and *Indian Chief;* the schooners *Randolph* and *Fidelity.* They were farmers and blacksmiths, coopers and sailmakers, barbers and carpenters and wheelwrights, their children and their wives.

Robert E. Lee, the American Confederate Civil War general, freed most of his slaves before the war and offered to pay their expenses to Liberia. In November 1853, Lee's former slaves William and Rosabella Burke and their four children sailed on the *Banshee,* which left Baltimore with 261 emigrants. Five years later, Burke wrote, "Persons coming to Africa should expect to go through many hardships, such as are common to the first settlement in any new country. I expected it, and was not disappointed or discouraged at any thing I met with; and so far from being dissatisfied with the country, I bless the Lord that ever my lot was cast in this part of the earth."

By 1867, the ACS had sent more than 13,000 emigrants to Liberia. But the settlers' death rate from malaria and yellow fever was staggeringly high. According to census documents, only 8 of the 86 emigrants who arrived in Sherbo Island in 1820 were still alive by 1843. Only 171 of the 639 who arrived in 1833 were still alive a decade later. During these years fresh graves dotted the settlements, which gradually expanded along the coast and up along the Saint Paul River.

Not everyone came willingly. Some historians suggest as many as 70 percent of African Americans sent to Liberia were told they would be freed *only* if they "agreed" to "go back" to Africa. And there was another group streaming into the new country under less than voluntary means. These were the "recaptives"—men, women, and children who had been snatched or sold into slavery but then rescued mid-Atlantic from slaving ships. Having abolished the slave trade throughout the empire in 1807, Britain sought to compel other nations to follow suit, using its mighty navy to enforce an embargo of slave trading throughout the Middle Passage. The United States, which abolished the slave trade in 1808 (though, of course, not slavery itself), also joined in the naval barricade.

Many of these people—Ibo and Fulani, Kongo and Yoruba, Bantu and Fon—had been taken from points hundreds of miles south and west of Liberia, from parts of what are now Benin, Nigeria, Cameroon, Equatorial Guinea, and Congo. To return them to their respective homes was impossible, so instead they were "liberated" in Liberia or Freetown. The American settlers dubbed this new group "Congo People," because they were assumed to have come from areas in and around the Congo River basin. Hastily enacted laws forced many into long apprenticeships with the settlers. Such apprenticeships were intended to civilize and Christianize the Africans, who, unlike the local indigenous peoples, were cut off from their cultural and tribal roots and thus could be more easily assimilated into settler society.

In 1838, the colonies established by Virginia and Pennsylvania merged with the ACS colonies, declared themselves the Commonwealth of Liberia, and claimed control over all settlements between the Cestos River and Cape Mount. After Liberia declared its independence in 1847, Joseph J. Roberts, a freeborn black born in the American state of Virginia, was elected Liberia's first president. For many, many years every president of the country would come from Montserado County, most of them from Monrovia: the City on the Hill.

My parents settled in a house on Benson Street, one of the city's primary thoroughfares. Their first child was a son, named Charles after Charles Dunbar. Next came a daughter, named Jennie after our grandmother. I was third, named after a friend of my mother, and then another son, named Carney after our father.

Ours was a neighborhood full of families and children, more of an extended village than a real city block. There were settler families and Gola families and Mandingo families and many other folk. There were teachers and shopkeepers and politicians too. The women cooked and went to market, the children played and did chores and went to school. On Sundays my mother would bake cornbread, sending children around to the neighbors to borrow any missing ingredients.

Our house was grander than some, more modest than others, a two-story concrete structure with coconut trees growing in the yard. We drew our water from a well in the backyard and, in the early days, did our business in an outhouse behind the yard next to a small stand of coconut trees. I was something of a tomboy growing up and loved to climb that coconut tree and to play football with the boys in a nearby field. The balls we used were no more than old tennis balls discarded by one of the few foreigners who had visited town and who had such things.

The toilet itself, with its rough plank boards stationed around and above the hole in the ground, was less pleasant—especially the time I fell inside. I was so small at the time, perhaps three or four, and someone had taken me out to sit me on the boards and left me there. When I fell in, I screamed and screamed, and my poor mother came flying from the house calling "My baby! Save my baby!" By the time she reached me, a passing neighbor had already pulled me out, and together they washed me off.

Possibly this was one of those times during my childhood when my sister Jennie took the opportunity to recall the old man's predictions of greatness.

Later on, as our family's fortunes increased, we built an inside toilet, becoming one of the few families in the area to enjoy that luxury. But one of the men who saved me from the outhouse is still in Liberia today, and every now and then he likes to remind me of his heroic act.

Once established in his law practice, my father entered political life. In short order he was elected to the Liberian House of Representatives, the first indigenous man to be so. This was a major accomplishment, made possible in part by William V. S. Tubman's election to president in 1943.

My father's dream was to become the first native Speaker of the House. For a time it looked as though he might accomplish his ambitious goal. He was well thought of by President Tubman, who not only appointed my father to many foreign delegations but often visited our house.

Naturally it was a big deal when the president visited. The house would be abuzz all day with preparations: cleaning and arranging and cooking great Liberian food such as *fufu* (dumpling) with palm butter, meat stew, country chops, *jollof* rice. When the president arrived with all his government officials, we children would be ushered into the background and warned to stay out of sight. But while the men sat on the porch eating and talking and laughing, we would peep around the corner to listen to their deep voices boom.

Outside work my father was a worldly man who spent much of his personal time away from the family and out on the town. He was good-looking and quite stylish, full of vigor and vim, and he had many, many friends. He also chased women with great enthusiasm and energy. This was not unusual nor particularly frowned upon at the time. Polygamy was the dominant form of marriage in Liberia, and most of West Africa, before the arrival of settlers and colonists. Even the settlers, who brought with them Christianity and its emphasis on monogamy, took up the practice of having concubines and "outside" children.

My mother, a devout Christian, was no doubt pained by my father's wanderings, but she kept any strain it caused away from us children. She took her solace in religion. Then she went out and created, as best she could, her own life.

She opened a school about a mile from our house, which we and dozens of other children attended throughout our primary years. Also, she became an itinerant minister in the Presbyterian Church, traveling throughout the countryside to preach the Word of God. It was rare in those days for a woman to serve as a traveling pastor, but my mother did. Sometimes my siblings and I went along on her missions, traveling hours by foot or by canoe to some small village or town for Sunday services. My mother had a strong and lovely singing voice, and she often used it in the pulpit to praise the Lord. "A Mighty Fortress Is Our God" was one of her favorite hymns.

It was on one of these trips that I got my first taste of public speaking. One weekend when I was about eight years old, my mother took my sister and me along on a journey to Careysburg, a town about fifteen miles northeast of Monrovia. We were to visit the local Presbyterian church that Sunday, and I was chosen to give the recitation. I spent the whole of Saturday afternoon sitting in a guava tree at the Ureys' home, studying my lines. By the time I went to bed Saturday night, I thought I had it. The next morning we rose and dressed and went to church, but when I was called before the congregation, I froze. I stood there at the front of the church for agonizing moments, trying to remember my lines, but could not remember a word. All I could do was stand there and look into my mother's pleading face. After a while they clapped me down and I slunk back to my seat in tears.

My mother, who had such an understanding heart, consoled me as best she could, saying, "Well baby, these things happen to people." Since that day I've given far too many speeches to remember, but I've never been unduly nervous again. I think something in my eight-year-old self vowed that embarrassing morning that I would never fail at public speaking again.

My mother spent some time with her real mother, Juah, in their village, but mostly with her foster mother, Mrs. Dunbar, in Monrovia. Grandma Juah also came to Monrovia often. Family connections are vitally important to Liberians and to Africans in general, and my mother was no exception. Grandma Cecilia, as we called her, had been vitally important to my mother, and she became vitally important to us children as well. Every Sunday after church we'd go to her house to spend the afternoon eating her delicious gingerbread and reveling in her love.

Above all else Grandma Cecilia was a lady, with all that the word implies, and she wanted us to be ladies and gentlemen too. All that she had given my mother she also gave to us, teaching us how to dress, how to speak properly and politely, how to eat at the table with good manners and dignity. So close did we remain to Grandma Cecilia that our oldest brother, Charles, was actually raised as much by her as by my mother.

When school closed for vacation, my mother, my siblings, and I would leave Monrovia and travel upcountry to spend the better part of vacation with our paternal grandmother, Jenneh. This area is very different from the coastal area, upon which most of our major cities stand. Along the coast the land is low, sandy, and flat, the shoreline broken by rocky capes, river estuaries, and tidal creeks.

But behind this low coastal plain, which extends for some thirty miles, the land changes. Up rise a series of low-lying hills and rolling plateaus bisected by rivers. Up springs the rain forest, the wettest and thickest in West Africa. Upcountry, we call it, or the bush.

This area is part of the Guinean Forest of West Africa, one of twenty-five biodiversity hotspots in the world. A hotspot is an area that contains large numbers of endemic species that are under severe threat of extinction.

The Guinean Forest Hotspot, in which are found some 9,000 species of plants, some 320 species of mammals, 785 species of birds, and 210 species of reptiles, covers portions of eleven countries in the region, but only Liberia lies entirely within the zone. More than 40 percent of the original forest cover survives in Liberia, and this area is home to, among many others, the pygmy hippo, an animal known exclusively in Liberia. It is also among the highest-priority regions in the world for primate conservation.

In Liberia the rainy season historically begins in late April or early May and lasts until October or November, deluging the country with as many as 200 inches of rain in some parts. There is no rain like West African rain. I have known it to shower water from the sky for as many as twelve hours straight, then pause and rain some more.

The dry season normally stretches from November until April, and during that time the land turns dusty and hot. During these months we return to school because the roads are passable. However, climate change is real. We in Liberia are already beginning to see and to feel its effects, as the rains seem to come earlier, fall heavier, and last longer than before.

Julejuah is a small village of several hundred people located some twenty miles from Monrovia as the crow flies. But there were few cars in Monrovia at the time and even fewer paved roads. To get from point A to point B, most people had two choices: to canoe one of the country's many rivers or walk. To get to Julejuah, we took a car partway, and when the road ended, we walked.

As the wife of a Big Man, not only the son of the chief but now an important government man, my mother could have spared herself the long walk to our father's upcountry village and been carried in a hammock. But she always refused; forcing men to carry other adults was demeaning, she said. "I will not make beasts of burden of them." She would, however, allow the men to carry us children, so as to avoid making a long trip interminable.

We would spend two and a half months in the village, and village life was wonderful for myself and my siblings. It was in the village that I learned to swim, in the clear, warm waters of the river that separated the village itself from the farm. In the early mornings we would rise to the sound of our grandmother, Nanna, who spoke very little English, calling us for breakfast: "Jennie, Ellen, Carney, Charley, come eat rice!" We'd all go scrambling from our sleeping area in her *rondaval*, a circular, mud-and-wattle, thatched-roof hut, and into the little kitchen she had built on the side. Nanna would be standing over a fire made of sticks burning between large stones which had been collected by the village men, cooking rice in a pot. After breakfast we'd race down to the river to swim and dive down to the sandy bottom or to work the canoe that took people across from the village to the farm. Later we would cross the river to join our grandmother and the other women on the farm to sit in the hot sun with branches and drive the birds away from the growing rice.

It was in the village, too, that we learned about Poro and Sandee, the traditional male and female secret societies whose rites initiate boys and girls into manhood and womanhood. When time came for them to take the children, there would be a lot of hollering and we would all be afraid. We would run under the beds and hide because people said the devil was coming to eat people. When the children disappeared, people said they had been swallowed by the devil. Of course, they had simply been taken into the bush for their initiation ceremonies, but we didn't know that. We were terrified.

Both my mother and father had been raised as Presbyterians, but many of the people in my father's village were Muslim, including my father's brother, Zwannah Ginde, a local Paramount chief, who introduced us to accepting the Muslim faith as part of our national culture. There was never any strife about this; in the countryside, especially, Muslims, Christians, and practitioners of tribal religions generally coexisted without tension or grief. When our Muslim friends stopped in the middle of the day to pray toward Mecca, we would stand respectfully aside. And when we said grace over our meals or celebrated Christmas, they would do the same.

My father remained in town for most of our vacation, but he would make the trip out sometimes, for an extended weekend or for the Christmas holidays. I can picture him now, standing tall in the canoe as it crossed the river, his loaded rifle on his shoulder as he took aim at a fat pigeon or a monkey. It was a big deal when he came; all the villagers would rush out to greet him and show their respect. He would hold court in the palaver hut, speaking to each villager in his native Gola dialect.

By spending time in the village, we children too picked up a few words of Gola, so that we could communicate with our grandmother. Many years later, at one of the most frightening moments of my life, those few words of Gola would save me from a terrible fate.

It was important to my father that we children remain connected to our indigenous roots, and for this I am grateful, because it might so easily have been otherwise. My father might have sought to disappear into the elite settler class, which had, on the surface, accepted him, taking upon himself the settlers' disdain for country people and tribal languages. He might have played the city life and embraced Americo-Liberian hegemony. He might have taken the hand of my mother—

who was very fair and did not much look like an African anyway—and together they might have simply walked away from their rural backgrounds. Many did. My parents did not.

So, during the 2005 campaign for president, when my detractors began implying that I was descended from the settler class, that I had no understanding of the hard and difficult lives of indigenous Liberians, I had to set the record straight.

I told them I knew what it was to be a part of a family denied opportunity. My mother and father had been blessed to have broken through the barriers, but when life circumstances struck our family hard, those barriers had risen again. Our gift, our strength, lay in our having always been mindful that such barriers still existed and that others were struggling mightily against them, because our grandmothers were always part of our lives. Neither one of my grandmothers could read or write any language—as more than three quarters of our people still cannot today—but they worked hard, they loved their country, they loved their families, and they believed in education. They inspired me then, and their memory motivates me now to serve my people, to sacrifice for the world and honestly serve humanity. I cannot and will not betray their trust.

In truth, I stand with a foot in each world. I represent both the good and the bad of our society. I represent both those who were given opportunities and advantages because of their background and those who were denied those very things because of who they were and from whence they came. My feet are in two worlds—the world of poor rural women with no respite from hardship and the world of accomplished Liberian professionals, for whom the United States is a second and beloved home. I draw strength from both.

In the end I have been, throughout my life, first and foremost, ahead of everything, a Liberian.

Everything else came afterward.

CHAPTER 2

# CHILDHOOD ENDS

MINE, THEN, was a happy childhood, rich and easy in the things that matter: health and safety, love and family, a sense of stability and of security. At least it was until the day my father suffered a massive stroke.

He was in his early forties when it happened, the prime of life, and he was still vigorously pursuing his goal of becoming Speaker of the House. For a time, it looked as though he might succeed. Under President Tubman, the doors to government service and political accomplishment were being opened to native Liberians, if only a crack.

Since independence, political power in Liberia had been monopolized by a small, concentrated group of settler families based in and around Monrovia. In some ways, William Tubman was a part of this elite group. His mother, Elizabeth Rebecca Barnes Tubman, was born a slave in Georgia and emigrated to Liberia with her parents after the American Civil War. His father's father, William Tubman, arrived earlier, emigrating from the state of Maryland in 1837, after he and his parents were freed in their slave owner's will.

But the Tubmans were settled not by the American Colonization Society but by the rival Maryland State Colonization Society, and they put down roots not in the bustling capital of Monrovia but 250 miles down the coast at sleepy Cape Palmas.

From the beginning, the Maryland settlement at Cape Palmas had stood apart from the other colonies. When Monrovia and the Saint

Paul River colonies joined forces with the Bassa Cove settlement in 1838 as the commonwealth of Liberia, Maryland declined to come aboard. Four years later, when the only other holdout, the Mississippi colony in Sinoe County, was admitted into the commonwealth, Maryland still remained outside the fold. Even after Liberia became an independent republic in 1847, Maryland stood apart.

It wasn't until 1857 that Cape Palmas, facing threats from neighboring British and French colonies as well as attacks from local Grebo chiefs angered by the settlers' efforts to regulate trade, joined the republic as Maryland County.

Tubman, then, grew up in Cape Palmas, a decided member of the settler class but an outsider to the Monrovia power base. His father was a rock mason and stern Methodist preacher. He himself also trained as a lay preacher before studying law and fighting in the army. His speaking skills caught the attention of President Charles King, who backed him in his election to the Liberian Senate at the age of thirty-five. Later he became an associate justice of the Supreme Court.

Tubman, known far and wide as "Shad," won the presidency with a powerful combination of charm, charisma, and great political acumen. Immediately after taking office he launched a policy of "national unification" intended to integrate the country's two societies, natives and settlers. Tubman extended government representation to the hinterlands, granted women and ethnic Liberians universal suffrage, and brought indigenous people into politics in a way no president before had ever done. He appointed native men to positions of power in his administration, increased educational spending, and threw open the doors of the Executive Mansion to regular citizens, who might come and request an audience of "Uncle Shad," as he was affectionately known. Tubman also banned the use of the elitist term "Americo-Liberian," declaring "We are all of us Liberians."

But in reality the Unification Policy was largely symbolic; the ideal of equality it represented lacked both substance and girth. Although tribal people from the country's interior gained representation in the legislature, they remained vastly underrepresented; power and privilege continued to be monopolized by the settler groups. And Tubman's economic policies brought what a critical 1966 study by Northwestern University researchers labeled "growth without development," meaning that what wealth flowed into the country from the rapid development of the economy remained squarely in the pockets of a very few. Tubman and his allies lived in luxury—building luxurious homes and grand public buildings in the capital, donning top hats and tuxedos for formal state occasions, and traveling the world—while the majority of rural Liberians remained dirt poor.

Tubman also forcefully oppressed political opposition, punishing those who rose against him or his ruling True Whig Party with harshness. He cultivated an atmosphere of fear and intimidation through his Public Relations Officers, or PRO, program—sometimes known informally as "people reporting on others"—essentially an army of paid civil service spies who served as the eyes and ears of government. Those who fell afoul of the power structure were punished—swiftly and effectively.

Tubman also consolidated his grip on power by constructing both a vast security force and, most important, the most potent patronage network the country had ever seen. In some ways, Liberia struggles with the legacy of that patronage system even today.

Nonetheless, to my father, President Tubman was a man who opened the door. Then illness slammed that door shut again, with a suddenness and ferocity that shook us all to our core. One day our father was young, vibrant, and powerful; the next he was a shell of his former self.

No one could explain what had happened. In the Liberia of the 1950s, medical care was rudimentary. There was no such thing as an annual checkup, no such thing as screening for hypertension that might have warned my father so that he could have done something to prevent the stroke. No one really knew what caused a person to be walking around apparently healthy and hale one minute and the next unable to move his entire right side.

My father, casting around for his own explanation, came up with one grounded in indigenous cultural beliefs.

"I have been witched," he said, his speech slowed and thickened. "Someone has put juju on me."

Juju is West African voodoo, witchcraft, sorcery. Its origins date back thousands of years, its reach stretches across the continent, and a belief in its power and effectiveness remains strong among many Liberians of all classes, as would become violently and devastatingly evident to the world during the civil wars that roiled our land in the 1990s.

But my mother, a deeply Christian woman, rejected all such belief and all such talk. She put the blame for my father's illness instead on my father's fast lifestyle. "You need to pray to God for healing," she urged him. "Pray for healing and for the forgiveness of your sins."

Still, my father—and many others—continued to believe he had been witched for his ambition and political drive, for daring to aspire to be the first native Speaker of the House. To my mother's dismay, he sent for various country doctors to come and reverse the illness. None of it helped, of course, and my mother grew exasperated. My father, I believe, grew increasingly depressed.

The stroke affected his right side, and although he eventually recovered enough muscle strength to be able to move around, it was always with great difficulty. He had to drag his foot along the floor as he moved. So too with his speech; right after the stroke his speech was thickened and slurred. As time passed his speech improved and we could understand him. But as more time passed and he remained ill, as he watched the family struggle and his attempts to reverse what he considered a curse failed, he sank more and more into himself. He spoke less and less as the years passed and spent more time sitting in his chair or in bed.

After my father's stroke our lives changed dramatically. Suddenly my energetic, on-the-go father was homebound. His colleagues vanished. His friends, the boys with whom he had spent so many fun-filled hours out on the town, trickled away. The family income plummeted. My mother began making baked goods to sell to make ends meet.

We stopped going to the village; the trip was too costly and far too difficult for my father to make. Although there were usually people around the house—friends, extended family, older children my mother had taken in—to help with his care, my mother shouldered the bulk of the burden, especially in those early years. In the best of circumstances, caring for a stroke victim is an intensive, exhausting, all-encompassing obligation. My father needed help with everything, every aspect of daily life. Every morning my mother would rise to bathe, dress, and feed him before helping him to the chair on the porch where he spent most of his time, staring out at the passing crowds. Later my sister, Jennie, would go off to England to become a nurse so that she could return and ease my mother's load.

By the time we all adjusted to this new life of ours, I was in high school at the College of West Africa. At the time, CWA, a Methodist institution, was considered the most prestigious (and was the oldest) high school in Liberia, a school that only the privileged and well-connected could attend. My father's position in the legislature had provided me that opportunity, but, with his sickness, time was running out. Still, at CWA I found a kind of emotional refuge among the dedicated, disciplined teachers and lively classmates.

I was also something of a tomboy and loved playing sports of all kinds. Sometimes I even played football with the boys, not a thing normally done by young ladies, but my favorite sports were table tennis and volleyball. Volleyball was my best sport; we played it every afternoon in the school yard. I was a serious and assertive player, and helped lead our school team to victory against rival teams.

The only problem at school was that some of my classmates teased me about the fairness of my complexion. They said I was too light to be a real African and called me Red Pumpkin, a name that hurt me to the bottom of my soul. Many days after school I cried my way home. Many nights I went to bed praying to God to let me wake up black. It was the one wish I ever wanted in all my life, to wake up black. It never happened, of course, and my mother would tell me to stop wasting time and energy regretting things that could not be helped.

It was while I was in the last year of high school that I met the man who would become my husband. James Sirleaf was the son of a

Mandingo father and a mother from a prominent and established settler family, the Coopers. In those days it was unusual for a member of the settler class to marry a Mandingo person, and there is little doubt that James felt some sting of discrimination while growing up.

The Mandingos are Muslims whom most historians believe immigrated into Liberia as early as the seventeenth century. They came primarily as traders and, because of their religion and lack of participation in certain central institutions, such as the Poro and Sandee societies, remained culturally, and sometimes geographically, apart from mainstream society. Even today some in Liberia, wrongly and unfairly, label Mandingo people "foreigners"—yet another example of the internal division we are all striving mightily to eliminate in Liberia. James's father died when he was young, and he grew up as a part of the elite Cooper family. But his Mandingo heritage was always there in the background.

We met at the movies. My friend Clave was invited to a show by a young man, who said she should bring me along and he would bring along his friend "Doc" Sirleaf. Everyone knew Doc (whose given name was James), and everyone knew he had just recently returned from attending college at the famous Tuskegee Institute in Alabama. What neither Clave nor I knew at the time was that, apparently, Doc had his eye on me.

Well, that was it. Doc was older than I by seven years, tall and articulate, ambitious and charming and determined to succeed, with a military bearing from his days in the ROTC. All of this lent Doc the scent of sophistication and adventure. Taken together, he was a man hard to resist.

We began to see each other. Very soon it emerged that he was the jealous type; he did not want me seeing other beaus, and so we soon began to date exclusively. He took me to my senior prom and to other functions. As Clave says, Doc swept me off my feet.

But our closeness rang alarm bells with my mother, who pulled me aside and warned me against temptation. "I don't want you to do bad things," she said. To Doc she insisted, "I raised my children to be upright."

My father, too, though disabled, still insisted on exercising his fatherly duties. When Doc came to the house to pick me up, my father called him in and demanded to know his intentions in spending so much time with his young daughter. The message was soon clear to both of us—if you are going to be involved, you had better be married. Anything else was scandal and shame.

I was deeply in love and wanted very much to be with Doc. But also I knew my prospects for college were not as bright as they might have been before my father's illness. I had come to understand the harsh realities of life. My more affluent women friends were going off to college abroad because their families remained privileged and connected. My father had gained a toehold into the settler class only by adoption. Now that he was no longer able to protect us, we were forced to fend for ourselves.

Seeing no other opportunities, I gave in to tradition and convinced my mother that I wanted to be married. Not long after I graduated from high school in 1956, we were. My brother walked me down the aisle in place of my father, who was seated in the front pew. It was a bittersweet moment for all involved, but the local newspapers had a good time with it. They called it a *tom-tom*, or small, wedding, because the bride, and half the wedding party, were so young. Clave was my maid of honor. We were both seventeen.

After the wedding Doc and I went off to his family's farm for our honeymoon. That Sunday, as was tradition, Clave and the rest of the wedding party came up for a visit. It was a strange and interesting thing to be standing with Clave, me suddenly on one side of the adult line, her still firmly on the other.

My sister, Jennie, was still in England at the time, training to be a nurse, and so missed the wedding. To say she was less than pleased about the event would be an understatement—she was upset. It was nothing against Doc; she simply felt I was far too young to be married. She feared my ambition and drive would be quickly squashed beneath the requirements of good Liberian wifehood. And sure enough, not long after my wedding I was pregnant with my first child. My son James (nicknamed Jes) was born on January 11, 1957, my mother's birthday. That December 31, my second son, Charles, was born.

Two children in one year, while my friends went off to college abroad.

For a while it looked as though my sister had been right.

WHEN DOC and I first met, he had just returned from studying agriculture at Tuskegee Institute. It would be difficult to overestimate the importance of agriculture to Liberia and Liberians. Since the founding of the country, farming has been the primary livelihood for the vast majority of Liberians.

Before the civil war, agriculture accounted for about 40 percent of GDP and Liberia was a producer and exporter of raw materials: iron ore, timber, and rubber. In Liberia, rubber is king. In the past, the rubber industry generated as much as U.S. $100 million in export earnings annually. In 2000, agriculture and forestry contributed more than 90 percent of export earnings, mainly from rubber and timber, plus cocoa and coffee.

When we took office in January 2006, we inherited an agricultural sector devastated by the civil conflict. Agriculture, being labor intensive, traditionally absorbed more than 70 percent of our nation's workforce. Today it remains perhaps the most vital sector of our economy and the prime pillar of our strategy for poverty reduction, food security, and economic recovery.

In order to start on the long road toward the achievement of this objective, we distributed more than 40,000 tools and 20 metric tons of rice seeds to some 33,000 farmers throughout the country. We also reactivated activities at the Central Agricultural Research Institute in Bong County and have completed a plan of action for its complete rehabilitation. In spite of this progress, the agriculture sector in Liberia faces continuing challenges.

For decades we Liberians have boasted of our rich and bountiful soil, and this is true—to some extent. Our fertile rain forests grow some 250 species of hardwoods, the majority of them marketable, and our cultivated farms, both small and large, produce cassava, rice, plantain, and other staples of the Liberian diet. But our heavy rainfall also causes leaching of nutrients and erodes the topsoil.

What's more, aside from the foreign-owned rubber, palm, and coffee plantations and some medium-sized farms that produce coffee, cocoa, sugarcane, and oil palm for cash, most farming in Liberia is subsistence, and most farmers are small. Practices remain low tech and traditional: farmers lightly clear small sections of land using hand tools, plant cassava, rice, and okra in mixed patches at the start of the rainy season, and then wait. Cultivation is shifting, livestock production is low, and most farmers grow only enough to feed their families or villages. Right now, domestic rice production meets only a small portion of the country's needs. The rest is imported, principally from Asia.

The wars had greatly damaged our agricultural sector, as farms and plantations fell into ruins after being abandoned by investors or were actively stripped and destroyed by combatants. We have started the process of resuscitating these plantations and other farms. Our objective in this regard is to ensure that the plantations are operating again under sound interim managements until they are reclaimed by the original investors or given to new investors under a competitive process.

Doc's goal—prior to the wars—was to bring what he had learned at school in the United States back to Liberia and to join the Ministry of Agriculture. But achieving that goal took some time. For the first few years of our married life we lived with his mother while Doc got on his professional feet.

To make ends meet, I knew I had to get a job. Fortunately, I had taken typing in high school and so was able to secure a job as a secretary for the Stanley Engineering Company. From there I moved on to a job at the Elias Brothers' garage, where I worked as an assistant to the head accountant, an old French expatriate man who had lived in Liberia for a long time.

That job was the start of my career in finance. Growing up, all I had ever wanted was to become an English teacher like my mother; the idea of pursuing accounting or economics had never entered my mind. I took the bookkeeping job for purely practical reasons: not only did we need the money, but the owner of the garage would sometimes let me borrow a truck at night so I could drive to the beach, children in tow, and pull sand to build the blocks for our first house.

Still, once I took the position, I had stepped into a stream that would carry me along toward the future of my professional development. So often it is the small decisions in life that end up shaping our future the most.

After a while Doc got a teaching job at the Booker Washington Institute, a vocational high school founded in 1929 by the ACS and other missionary and philanthropic groups. Situated on a thousand-acre campus in Kakata, about forty miles outside Monrovia, the school had prepared generations of Liberians for jobs in carpentry, auto mechanics, masonry, and agriculture. We also spent a lot of time at the family-owned small farm several miles away, and there we set up house with the children. Although there is a lot of help, country living in Liberia is a tough life, one of nearly constant work: hauling water, building fires, planting and harvesting cassava and rice to eat. And all the while having more children. By the time Doc landed his desired job with the Ministry of Agriculture and we had returned to Monrovia, I had four children, all boys.

My sister had returned from England by then, married, and was raising her own family: three children, also all boys. We helped each other, she taking my sons while I tried to work at various little jobs, me taking her children when she went to care for my father or, later, to pursue her nursing career. It was an amusing sight, me moving around the streets of Monrovia with seven little boys around me. We had a small car by then, a Volkswagen Beetle, and I would pile all seven children into that car and head off to whatever I needed to do that day.

But I felt that something was missing, that this was not quite the life I was meant to have. All my friends who had gone off to college in Nigeria or London or the United States were returning now, sophisticated and professional. One by one I watched them mount the ladder of success and begin to climb. Meanwhile, I stood off to the side, mothering my children and working a series of low-paying, dead-end jobs that were taking me nowhere fast.

I looked around and saw the lives of so many Liberian women, all of these incredibly hardworking market women and housewives and mothers, and what I saw was that their lives were drudgery, a simple trudging from day to day to day. I did not want that; that was not the life for me. My own mother had struggled and beat the odds, clawing her way up from the unrelenting grind of rural African poverty, and I was not about to fall back into the hole. One thing I had always believed in was my own potential, and I knew it did not lie in filing papers and collecting payments as the secretary in an auto garage. Nor, as much as I loved my children, did it lie in simply raising them.

But at that point I could see no way out. We certainly didn't have enough money to even think about sending me away to school. So I kept working and raising the children and feeling depressed. I think the low point was one month when Clave, who had gone off to the United States to attend college, came home for vacation. She came to the house to visit me, and though she was lovely and nonjudgmental, I suddenly saw myself through her eyes: a struggling housewife with no future, just a bunch of children hanging on to me. Clave says now that she could see I had lost my childhood, but it wasn't my lost childhood I was worried about. It was my future that concerned me more.

This can't be the end for me, I thought. I have to get out of this. I have to do something.

When Doc was awarded a scholarship to pursue his master's in agriculture in the United States, I saw my chance and lunged at it. I applied for a government scholarship so that I could accompany my husband and further my own studies. Had my father been alive and still a member of the legislature, the application would have been a mere formality, the scholarship given without hesitation or note. That was the way things worked. My sister, Jennie, had received such a scholarship to study nursing in England, on both her own merit and the strength of my father's connections.

But things had changed for our family. My father, long abandoned by his legislative colleagues, had passed. My mother and I began begging for the scholarship and did not stop. I sat an exam and pleaded with anyone who would listen and might help, until finally the scholarship was granted. I was told I could go. Doc would be studying agriculture at the University of Wisconsin at Madison. I would study business at Madison Business College.

There was a reason to hesitate, of course—the children. My youngest son was barely a year old, the others not much more. Could we take them with us, all the way across the great Atlantic and into the unknown world of Madison, Wisconsin? No, it wasn't feasible, either financially or practically. Both Doc and I would be so busy studying and working that we would not have time to care for the boys, and with no family around in a foreign country we would have no reliable help.

There is much Liberia has to learn from the industrialized world of the West. But one thing the West can learn from Africa and the developing world, one aspect of traditional African culture I hope my countrymen will never lose, is the awesome support of the extended family system. In Liberia, as in all of Africa, parents routinely help their children care for the grandchildren—not just temporarily, not for a few hours or for the night but for weeks and months and even years. How routine is it for a Liberian woman to take her children to her parents and ask them to care for them because she has to work or go to school. No payment is given, no money left unless you happen to have it, which probably you do not. Whatever means the parents possess, however meager, are stretched to feed, shelter, and educate the grandchildren while the adult child goes off to improve herself.

We left our boys with our families: two with Doc's mother, two with mine. It was a heart-wrenching decision and a painful good-bye, especially with my youngest son. The separation created a hairline fracture in our relationship, the slightest of strains, that remains to this day. There was a time long ago, while we were talking and reminiscing about his childhood, that my youngest son paused me in a point and said, "But, Mom, you can't remember. You weren't there." Today, although we have become quite close, some guilt remains.

Doc and I flew to Madison in the fall of 1962. When I stepped off the plane that first crisp September morning, I was both heartsick at having left my children and excited beyond measure to be in America. Here, finally, was the land about which I had heard all my childhood, the place from which the founders of our nation had come, the place everyone in Liberia wanted to see. I was anxious, thrilled, and determined. I was also a little cold.

WE RENTED a small apartment on the second floor of a two-story house and started our classes. That was when the problems began.

It is not easy, living with a jealous man. From the time we had first begun dating, Doc had displayed a possessive streak. In those early, heady courting days he would take me to dances and social events and we had so much fun together. However, the minute I danced with someone else, the minute some high school classmate or friend of the family came along and paid too much attention to me, he would create a scene. Sometimes he would go so far as to leave the dance, go home, put on his full military uniform, and return to intimidate his "competition." I saw all this, but at first I paid little mind. I was young and in love and believed a man had the right to protect the virtue and reputation of the woman he loved.

Once we were married, his possessiveness only grew worse. It got to the point that I had to be careful of even smiling at anyone, because it might set his temper off. He could act so unpleasantly that people simply did not like to be around him, or us, and friends often drifted away. Only Clave refused to be driven from my side. She just ignored his hostilities.

"I'd just tell him, 'Fine, you're cracky, but I can be cracky too!'" Clave says. "But he ran everybody else off."

If there is about me today a certain reserve or guardedness, as I have sometimes been accused of, perhaps this is the source. Clave believes it is. I was never as exuberant and outgoing a personality as my sister, Jennie, but Clave says that had I not married such a man at so tender an age I might have become more open. Is this true? Who can say?

Nor did I realize, when Doc and I were dating, that my Prince Charming had a drinking problem. When he drank, he became more jealous, less disciplined. He became mean, angry, stubborn, and unsupportive. I remember one incident after we were married, when I was working in the garage. I took his car out one day to run some errands. Far from home the car broke down. I hitched a ride home to tell him what had happened and to ask for his help. Doc just looked at me and said, "You better go bring me my car."

Looking back, it's difficult to say whether this was simply his character or whether he believed he was acting in the way a good African man was supposed to, being in control, keeping a firm hand on his wife. He was a hardworking, exceedingly disciplined man; that was his military training, and he thought everyone should be equally disciplined. In any case, such incidents were painful for me. I thought it was wicked on his part, wicked and cruel. But living with a person like that also strengthened me. I have to thank him for that, because it was a strength I would need in the days and years to come. That day I left the house, hitched a ride to the garage, found a friend to come with me to fix the car, and then drove it home. I got that car home, and I learned a valuable lesson: when push comes to shove and at the end of the day, depend on yourself. No one else is guaranteed to be there—no one.

In Madison, Doc must have feared he was losing his control over his wife, because his jealousy worsened. I spent much of my nonclass and nonstudy time working; we were poor, struggling students and needed the extra cash, so I got a job sweeping floors and waiting tables at the Rennebohm Drug Store on the edge of the University of Wisconsin–Madison campus.

Rennebohm Drug Stores were Madison icons, famous for their grilled Danish, Bucky burgers, and phosphate drinks. The stores, which numbered more than two dozen across the city, boasted long lunch counters, pink plastic dishes, speckled red counter stools, and squeaky Formica booths. Oscar Rennebohm, the founder, had been a successful businessman and onetime governor of Wisconsin. Only two years earlier, students in Greensboro, North Carolina, had launched a

nationwide movement against segregated lunch counters and restau-
rants by sitting in at a Woolworth's counter in that southern city. But
in Madison such blatant discrimination did not exist. Though the city
no doubt had its share of residential and housing segregation, restau-
rants, schools, hospitals, and public transportation services were all integrated. Rennebohm's hired me right off the bat. I was happy to have
the job, not being ashamed to work for an honest living. But Doc felt
the job was demeaning, that the wife of a man of his stature should
not be sweeping floors. He thought that type of work reflected poorly
on him, and he was too ashamed to admit that we needed the money.
He also thought the time I spent working interfered with my ability to
take proper care of him.

One evening I went to work as usual and quickly got busy cleaning up
the place. The store was crowded, packed with laughing students drinking coffee and eating ice cream. I was sweeping the floor when the door
burst open and Doc stalked inside. He walked right up to me, ripped
the broom from my hands, and began yelling. "You should be home!"
he screamed. "I demand you leave this moment!" He tossed the broom
across the floor. Everyone in the store stopped to watch, as I stood there
embarrassed and horrified. Finally I decided the best thing was to end
the scene as quickly as possible by leaving with him, which I did.

The next day I came back to see my supervisor. She was an older
white woman, very kind, and she said, "You've got a problem, I see."

"Yes."

"You are a good worker, Ellen," she said to me. "If you want the job,
you can still have it. But you have to tell your husband he can't come
here ever again."

We shook hands, and I went home to point out to Doc how much
we needed the money the job was bringing in. He grumbled for a
while but did nothing. The next day I returned to work, and every day
after. I remained at the drugstore for year or so, eventually graduating
from cleaner to cashier. Doc would fuss, fume, and feud every time I
left the apartment, but he never caused another scene at the store. In
that regard, at least, he was a practical man.

In most ways, I enjoyed my time in Madison; the city was easily negotiable, and the people were kind. I developed a love of American football and spent some of my rare nonstudy time with newfound friends cheering the Green Bay Packers on to victory.

The only major drawback was the weather. The kind of cold I experienced in Wisconsin had been all but unimaginable to me beforehand. Growing up we had read, in our large supply of books about America, about things such as snow and ice, but these elements are difficult to envision in a country where it's hot and humid all the time.

In Madison I was so cold I sometimes feared my tears would freeze. After I left my job at the drugstore, I got one as a clerk at a Midas muffler shop. This job required that I walk to and from work—more than a mile each way. Walking home one time during a snowstorm, I became so chilled I thought I would lose my hands. I had gloves on, of course, but by the time I reached home my hands were so red and numb I had to soak them in warm water to bring the feeling back.

I was in Madison on that fateful day in November 1963 when America began to lose some of its innocence. I was at the drugstore, working, when all of a sudden a kind of wind of words seemed to sweep through the place: Kennedy has been shot! The place stopped cold and everyone froze for perhaps one or two painfully long minutes. Then, after that moment of complete and utter silence, someone began to cry. Others joined in, sobbing. A few people raced out of the store and on to the street in their pain and bewilderment. It was bedlam. I stood where I was, too stupefied to move, thinking "How can they do this? How can they kill someone so good for the country?" For us, for Liberians and Africans and people all over the world, President Kennedy had been a great person, a powerful and glamorous figure, an idol almost. It was astonishing to think that someone had taken his life.

Doc remained in Madison for only the year it took to complete his master's degree, but in that time things went from bad to worse with us. The verbal abuse escalated surely and dangerously into the physical. One time we had some friends from home visiting the city and I went out to dinner with them and stayed too late. When I came home Doc was there, furious that I had neglected to make his dinner. He pulled out his gun—he always had a gun, considered it part of his necessary equipment as a military man—and struck me on the head with the butt of it. It wasn't a hard strike; he could certainly have hit me harder. Still, it hurt sufficiently. Doc always did enough to hurt but not enough to maim or kill. Just enough to keep me in a state of fear.

As Clave says, he had a violent spirit. Where it came from and why, I do not know. Perhaps it came from his mother, who had to survive and care for her children all alone after their father died.

Domestic violence knows no geographical boundaries. It exists in every nation, every society, every corner of the world, and neither Africa in general, nor Liberia in particular, is immune to this particular disease. Right from the start of my administration, even in my inaugural address, I pledged to bring the full weight of the government against those who would continue this terrible abuse. "Those who violate our women and girls now know they will bear the force of the law," I said. We have begun to do just that.

At the time though, the late 1950s and early 1960s, women simply took the occasional slap, or even worse, as a part of life. There was no recourse. There was no one to complain to, no police to seek or court administrator to whom you might appeal. The only thing you could do when you were beaten—and I did this more than once—was to run from home and straight to your mother's house. Your mother, though, was equally helpless. What could she do? Not much. Perhaps she would soothe you, tend your bruises, stroke your forehead. Maybe take you back home after a few hours, try to talk to your husband, do a little bit of admonishing. She might call a family meeting, appeal to the husband to stop such foolishness. Sometimes she might even try appealing to his mother to intervene. But in the end she would leave and the husband would either change or, more likely, beat you harder next time. What else could you do? This was simply part of life.

In my case, Doc's increasing verbal and physical abuse led me to ask myself a simple question: How can I get out of this life?

Doc finished his degree in a year and left the United States to return home. I stayed on for another twelve months, working and finishing my degree. When I returned home, we collected the children—who had grown so much I could not believe it—and resumed our family life. Doc took up his slot at the agricultural ministry, while I took a position as the head of a division at the Treasury Department. This was, in real terms, the beginning of my professional life. I was on my way, finally.

All the ambition that I had tamped down during those early days of marriage and motherhood came roaring to the surface. At the Treasury Department I had found my niche. At the department my position as chief of the Debt Service Division focused on domestic debt and debt relief.

With my mother and my sister pitching in to help with the children, I was able to pour myself into my work. My hours at the office stretched longer and longer, and what work I didn't finish there I often brought home. At night, after all my domestic chores were completed and we had all gone to bed, I would often slip from my bed and sneak out to the dining room table with my adding machine or my books. I was forever working and studying, determined to advance my career. Plus, I loved it.

But my husband did not, and the harder I worked, the higher I climbed, the more unhappy he became. The unhappier he became, the more he fussed and the more we fought. Sometimes, noticing my absence from bed at night, he would rise and storm out to the dining room to check on me. "What are you doing?" he would demand, standing before me. "At this hour you should be sleeping!" Another scene would inevitably follow, with much yelling and the not-infrequent slap.

Doc was also engaging in extramarital relationships. Eventually, the tension, fear, and abuse in our marriage drove me to do the same with a good friend, a man I came to lean upon during those difficult times and with whom I had a lifetime relationship ended only by death. None of this helped, of course; involving other parties in a troubled marriage just makes things worse for all involved. Day after day, month after month, the strain increased. Our marriage began to crack.

Doc began to drink more heavily, disappearing for hours and then coming home drunk. Sometimes he would remain in the car that way, just sitting there for hours while we watched from inside the house. I never minded when he worked off his feelings that way, because the alternative could be frightening. Twice he took a gun and actually put it to my head. "Move," he said, "and I'll blow your head off."

One day Clave, who knew all about the physical abuse, who saw it or the aftermath with her own eyes, looked at me and said, "Ellen, he's going to kill you if this keeps up. You have to let him go."

She was right, of course. Perhaps the worst incident, the one that finally cinched my dawning realization that I would have to leave, involved my first son. He was eight years old or so, just beginning to leave childhood behind. I had come home from somewhere and Doc was waiting, drunk and furious. He had a gun, and when I came into the house he pointed it at me. My son, who was in the room watching the events unfold, rushed out and returned with a can of mosquito spray, which he pointed at his father's eyes. "Stop!" he screamed, pushing the button. He was not tall enough to reach his target, but he tried nonetheless. Doc looked at our son in shock and put the gun down. It sobered him, at least for the moment. And it sobered me as well. I knew that as difficult as it would be—and it would be difficult—I had to leave.

Like nearly everywhere else in the world at the time, Liberia was very much a male-dominated society. Though Liberian women had long worked outside the home and even held positions of prominence in the government, and although African women in general are honored as mothers and aunts, women were not regarded as equals. In culture and practice, in spirit and in law, men were heavily and blatantly favored. I knew that if I walked away, or even if Doc walked away, he would be the one to get custody of the children. This was traditional in our society—children belong to the father first and foremost.

And it came to that. When it finally became clear that we could no longer live with each other, Doc informed me that he would be taking our sons. "You can leave and go to your mother," he said. "I'm going to take our children." So once again I was separated from my sons, and the pain was intense.

Even after the separation I was anxious about actually seeking a divorce. I wasn't sure what Doc would do. When it was time to go to court, Clave went along to be my witness and stand with me. By this time I had climbed high enough in the government to make connections, and I was known by many people, as was Doc. Everyone in the courtroom was nervous about what he might do if he appeared. Clave, who held the money to pay the lawyer in her pocket, kept urging the lawyer to hurry up and finish, threatening not to pay him if he did not. Even as the judge called the case, everyone kept looking over their shoulders, waiting for Doc to come storming into the room in his full military uniform. I think that had he appeared the entire court would have fainted, or perhaps run away. The case would never have gone through. However, he was out of town and did not appear to contest the suit, and so the divorce was granted. I was free.

For some time afterwards, as I continued rising in the Treasury Department, Doc would sometimes come by the office when he had been drinking and make a scene. That all ended after he remarried and later moved to Florida with our youngest son and we were able to become friends again. When he died, he was, as per his request, cremated, something unusual for a Liberian, especially at that time. I gave the keynote statement at his funeral.

After the divorce our two oldest boys were sent to a mission school, while the two youngest remained with Doc for a time. After a while, though, our youngest son, Adamah, went to live with his father's brother Varsay in Yekepa, Nimba County. This uncle was a medical doctor who worked in the mines; perhaps this helps explain why that youngest son is today himself a physician.

The third child, Rob, simply refused to stay with Varsay or with his father. He would cry and plead and demand to go to his mother. He agitated so fiercely and so relentlessly that finally one day Doc brought him and his bags and dropped them onto the step. "Here's your son," he said. "Take him."

And so it was that Rob was with me on the next step of my journey, my trip back to America.

## CHAPTER 3

# AMERICA AGAIN

DIVORCE IS difficult, even when it is absolutely necessary. Though I knew what I had done and what I must do in ending my marriage, still I felt, for a while, set apart. Whenever I had to fill out a form of some kind—and suddenly they seemed legion—upon which I had to indicate my marital status, a certain sensitivity would creep into my mind. Would I be marked as one of those people who had done something bad in a marriage? One of those people who simply could not manage a husband and a long-term commitment of that kind? It was disconcerting, but there was nothing to do but go through it.

My family was nothing but supportive. My sister offered encouragement and my brothers support. My mother took me in for a time, both literally and figuratively, praying all the while. And so I began moving on with my life.

From my spot at the Treasury Department I had a clear and unimpeded view of our nation's economy, and the view wasn't rosy.

Liberia's economic status had been precarious right from the founding of the nation-state. The ACS was a private organization dependent upon fund-raising to support its African ward. Once the settlers reached Africa, support from the society was minimal at best, and it became even more irregular once independence was declared.

The ACS managers continued to encourage agriculture as a way for the colony to achieve self-sufficiency, and by the 1860s a few upriver planters were successfully growing and exporting coffee and sugarcane. Between 1847 and the early 1920s, the production of these crops, plus ivory, palm kernels, and piassava reached an all-time high, and Liberia became an exporter of those products.

But fierce international competition crippled Liberian exports toward the close of the century, triggering a ripple effect. A decline in trade meant trouble for the country's merchant traders, who could not maintain their ships and crews. Fewer exports also meant lower customs duties flowing into the government's treasury, which led to trouble repaying foreign loans. Settlers abandoned farming and began what would become a long reliance on canned and salted products imported from the United States to meet domestic food requirements. And ambitious Liberians dropped agriculture and business as a primary vehicle for building wealth and turned instead to politics.

By the time Charles Dunbar Burgess King assumed the Liberian presidency in 1920, the nation's international debt stood at nearly $2 million, with American and European creditors knocking loudly and threateningly at the door. In his inaugural speech in 1924, President King was forced to ask, "Why, then, do we Liberians continue to live in Liberia and board in Europe and America? Such a course is sapping the vitals of our national life and rendering our boasted independence a mere shame and disgrace. For a country of its size and age to export practically nothing in the life of cultivated crops or domesticated animal products is a serious reflection."

Then too, as late as the 1930s, some leaders of the elite settler class were still counting on an exodus of black Americans to shore up the country and its economy (though this sentiment would cool somewhat by the 1960s, when President Tubman would establish a special commission to oversee, and limit, Black Power and Black Muslim immigrants from the United States and the Caribbean). But when a wave of black Americans finally did leave the American South in search of greener pastures during the Great Migration in the 1920s, '30s, and '40s, it was to industrial northern U.S. cities they fled, not across the ocean to Africa. Liberia continued to struggle economically.

Then came World War I—and, in Liberia, Firestone.

One cannot talk long about Liberia without discussing the Firestone Tire & Rubber Company. It is our largest private employer and runs what is considered the largest rubber plantation in the world within our borders. For better or worse, no other single company has had a more significant impact on the history and development of our land.

Firestone arrived in Liberia in 1926, excited about the country's perfect conditions for growing rubber as an alternative to its single Asia source and intrigued by the small, defunct British commercial operation at Mount Barclay, a lowland coastal plantation of about two thousand acres situated on a former mangrove swamp and jungle about twenty miles east of Monrovia.

Firestone and the Liberian government easily reached an agreement for Firestone to take over that plantation; the company was granted a long-term lease for $1 an acre the first year and a flat $6,000 per year thereafter. But Firestone had larger plans. After much negotiation, Firestone was granted the right to lease up to 1 million acres of "suitable" Liberian land for 6 cents an acre and 1 percent of the tax value of the rubber exported—and to do so until the year 2025.

By any measure, it was a sweetheart deal for Firestone.

According to the Dutch economist and historian Fred P. M. van der Kraaij, after the draft concession agreement was approved by the national legislature, Firestone suddenly introduced a new clause. This so-called Clause K made the agreement dependent on a $5 million loan from Firestone to the Liberian government.

At the time of Firestone's establishment in Liberia, the nation's economy was stagnant and bankrupt. Although the loan proposal and ensuing negotiations raised fierce protests both outside and inside the country—where some Liberians feared the influence such a loan would create on the Liberian government—under pressure from the U.S. State Department and eager for the cash to repay a $1 million debt to British bankers, Liberian officials eventually agreed to the deal.

Thus Firestone gained—for nearly a hundred years—almost unlimited control over an area equal to 4 percent of Liberian land and nearly 10 percent of land considered arable. And, by virtue of the loan, the company's entry into Liberia served mainly to reinforce Liberia's financial dependency.

For the next eighty years Firestone amassed huge profits and had a strong and decisive say in Liberian politics. The Mount Barclay site—renamed Harbel for Firestone founder Harvey Firestone and his wife, Idabelle—grew into a town as large as nearly any other in Liberia.

By the 1960s, there were as many as 20,000 workers at Harbel; over the years that number dropped, until by 2005 there were some 14,000, of whom 6,000 were not full employees but contractors or casual laborers, with no attendant rights. Those tappers worked in what an international workers' rights group called, as late as 2006, a "gulag of misery," according to a story in the *Philadelphia Inquirer.* That group, the International Labor Rights Fund in Washington, D.C., has filed a federal lawsuit against Firestone alleging that the company overworks and underpays its workers while exposing them to hazardous chemicals, inadequate safety measures, and harsh living conditions.

Conditions at the plantation were brought to international attention in March 2005, when a Liberian grassroots organization called the Save My Future Foundation published a report detailing what it said were labor and environmental abuses, including fourteen-hour workdays and child labor.

According to the report, most of the children of the laborers at the plantation were not attending school because of the shortage of facilities and the need to help their parents make their quotas for the day. Those who were in school attended substandard facilities.

From the report: "There are forty-five (45) divisions in the company's concession with four (4) junior high schools; there is no high school attached to the Company's school system. Most of the present junior high schools were formerly elementary schools of two classrooms that have been attached with additional classroom, made of local materials, by parents to host their children. In most of the Company's camps, laborers live in single-room mud huts, with no electricity, access to safe drinking water, latrine and bathrooms."

Indeed, the situation regarding the welfare of workers at Harbel had been ignored by successive Liberian governments, allowing Firestone to keep them in virtual servitude. The tappers labored long days in difficult conditions for little pay, lived in windowless one-room shacks without electricity or toilets (though managers live in lovely, comfortably modern redbrick homes nearby, adjoining a staff club and nine-hole golf course), and received education that rose not far above elementary level, creating, in effect, generation upon generation of nothing more than rubber tappers. Except for the lucky few, there was no way to break free.

Nor did all those tappers go to work voluntarily for Firestone. In 1930, the company employed more than 10,000 laborers on its plantations, according to van der Kraaij, but perhaps as many as 8,500 had not come of their own free will. A 1930 report by the League of Nations found that forced labor, supported and endorsed by the Liberian government, including then-President Charles B. King, was supplying many workers to Firestone.

In the meantime, although Firestone contributed to the nation's infrastructure in ways that helped itself, it did comparatively little to help the country in general. Most major infrastructure projects in Liberia during this time were financed through U.S. government loans. They included construction of a road from Monrovia to Totota (the location of President Tubman's farm), expansion of Firestone's private airfield into a national airport (Roberts Field), and construction of a deepwater port at Monrovia. The latter two investments provided the United States with access to facilities important to possible intervention in conflicts in North Africa and the Middle East.

Meanwhile, Firestone collected resources but established no manufacturing plants in Liberia, though it did establish tire plants and other rubber-processing factories in other African countries, including some that are not rubber producers.

It is only through new negotiations in 2007 that Liberia will get a true value-added business from Firestone: a plant that processes the dead rubber trees into veneers. This is still far from the standard we want, but it is progress and will create more and better jobs for our workers. Also in the renegotiated agreement there is a firm five-year plan for Firestone to greatly improve housing, education, water, and security for its workers.

The infusion of monetary resources did help the struggling Liberia during the 1920s and early 1930s. However, with the coming of the worldwide economic depression in the 1930s, the nation was struggling again. Thus little of real substance took place in Liberia in terms of growth during this period, despite Firestone's presence. And the political crisis involving President King further aggravated Liberia's economic woes.

Then, in 1944, came the man known as "Uncle Shad."

President Tubman's Open Door Policy, announced at his inauguration in 1944, was essentially a plan for unrestricted foreign investment, investment that, Tubman said, would "strike the rock of our natural resources so that abundant revenues may gush forth." Foreign companies were invited to operate in the country free from restrictive rules and regulations and granted favorable terms along with an impressive array of tax deductions and exemptions.

The policy spurred the introduction of lumber milling, tree plantations, and especially the mining of iron ore in places such as Nimba County. Perhaps most crucially, the policy helped pave the way for foreign involvement in the nation. No fewer than twelve foreign concessions in iron ore, rubber, and timber were granted in short succession. Aggregate investment by these companies totaled more than $500 million.

Among these companies was the Liberian Mining Company, founded by an American former lieutenant colonel named Lansdell K. Christie. According to a story in the March 28, 1949, edition of *Time* magazine, Christie had made a small fortune operating a barge line in New York. After being discharged from the U.S. Army, he came to Liberia, with which he had become familiar while helping the Army build Roberts Field airport, and wrangled a lucrative concession from the Liberian government to mine the rich iron deposits in Bomi Hills.

The deposits in Bomi Hills were of an extremely high quality, with an iron content of some 64 to 68 percent. Christie recognized these "mountains of riches" for what they were and quickly began exploration and development of Bomi Hills.

According to van der Kraaij and Liberian government records, Christie received an eighty-year concession giving him exclusive exploration rights to about 3 million acres around the Bomi Hills and exclusive mining rights for a smaller area. In exchange, the government received, according to van der Kraaij, "the promise of an exploration tax of US $100 per month and, after the 3½ year exploration period, US $250 per month. Further, a surface tax of initially 5 cents per acre and which would gradually climb up to 25 cents, and a basic royalty of 5 cents for each ton of iron ore exported. The concessionaire became exempted from all taxes in lieu of this exploration tax, surface tax and royalty."

Christie then persuaded the U.S.-based Republic Steel Corporation to come in on his deal, creating LMC. The company fared quite well during its lifetime. According to van der Kraaij, between 1951 and 1977, LMC shipped $540 million worth of iron ore. Out of that, the Liberian government received about US $75 million—just 16 percent of LMC's sales income. Until 1960, LMC's revenues surpassed the total revenues of the entire Liberian government.

In 1977, LMC closed its doors, having depleted the country's first iron ore mine. Liberians joked that Bomi Hills had become Bomi Holes, for indeed, after some thirty years of operation, all LMC left behind was a hole in the ground. Not even the primary facilities built to house and care for workers lasted much beyond the company. And although the main component necessary for road paving is crushed rock, of which LMC had plenty to spare, the road to the primary site remained unpaved for years, well into the Tolbert administration.

Then there was LAMCO: the Liberian American-Swedish Minerals Company. It was run largely by the Swedes, who built slightly better facilities than LMC had. Still, the uneven profit sharing between the foreign investors and the people of the land from which those mountains and mountains of profit came is something that leaves us all very sad.

Also started during this period was the Liberian maritime program, known as flags of convenience, begun in 1948 by the American Edward R. Stettinius, who had exerted significant influence over Liberia's economy while U.S. secretary of state during the Roosevelt and Truman administrations. After his retirement in 1947, he negotiated a major concession agreement with the Liberian government that created the Liberia Company and the Liberian Registry.

The group established the Registered Agent Office in New York City to register both corporations and ships under the Liberian flag. In this way, business owners could avoid certain U.S. taxes and restrictions. The program followed a similar one in Panama; after establishing it there, they established it here.

As with Firestone, the registry program brought in revenue that was critical to a country as small and struggling as Liberia. Little wonder, then, that no one in the Liberian government challenged the program or its management in any way.

In fact, the registry was essentially run from the United States, with Liberian officials relegated to receiving—and cashing—whatever checks "the agency" saw fit to send the government's way. It wasn't until Charles Taylor took power that some changes in the program were instituted—not so much to bring greater benefits to the country but to allow Mr. Taylor to better control the revenue himself.

(Interestingly enough, during the Taylor years, although U.S. sanctions were imposed on our timber and diamonds, no sanctions were ever imposed on our maritime program, even though it was known to be one source of funds used to purchase arms. The registry was one of the government's few legal sources of income during the Taylor years.)

At the time of this writing, my administration is at the beginning of renegotiating the terms of this important program. Whatever the results, we can say that today the Liberian International Ship & Corporate Registry, or LISCR, is the second largest fleet in the world, with more than two thousand ships representing some 10 percent of the world's oceangoing fleet. Some 40 percent of U.S. ships are registered under the Liberian flag. The registry is also one of the world's safest, as measured by underwriters and safety agencies throughout the world. This is to the credit of both the agency and Liberia itself. In that way, at least, what began with questionable motives has evolved into an enterprise in which Liberia can claim no small measure of pride.

However, there are changes that must be made. Over the years the registry's financial management has come under some criticism. What's more, as with Firestone, the question of whether the nation's legitimate interests have been truly served in profit division remains an open one.

We want to ensure that LISCR not only generates revenues for Liberia and that those revenues are properly accounted for and used in the best interests of Liberia and the Liberian people, but that the registry becomes a more fully involved participant in the country.

There is a strong need for better coordination between the registry and the Bureau of Maritime Affairs and Ministry of Finance. We are also calling for more registry training and employment for Liberians. There is no reason more Liberians should not be involved in the understanding and effective management of such a crucial part of our economic picture.

It is true, then, that the 1940s and '50s were a period of great growth for Liberia. In fact, the country registered, over the 1954–1960 period, growth in production and income averaging around 15 percent a year in nominal terms and 10 percent in real terms. Only Japan experienced greater growth during the same period.

However, there were serious structural deficiencies associated with the expansion, deficiencies that at the time went largely unnoticed and unaddressed. For one thing, most of the growth was concentrated in a small number of export-oriented, foreign-dominated firms whose profits flowed out of Liberia as swiftly as the natural resources upon which they were based. The situation was so advantageous for business that Firestone Liberia, in 1951, retained a profit that was three times the total income of the Liberian Treasury.

Also, with the notable exception of education (in which President Tubman aimed for a 50 percent literacy rate and a doubling of primary and secondary school enrollment by 1971), the government paid little attention to public sector activities. No one focused much on expanding the nation's productive capacity or diversifying the economy to create the structural changes necessary for sustainable growth. Liberia would live or die by rubber and iron ore.

In addition, few Liberians outside the settler elite truly benefited from the influx of foreign businesses. Thousands of Liberians were given jobs, but almost always lower-level, manual-labor positions, with little effort made to train indigenous workers so they might move up to technical or managerial slots. Hospitals and schools were built but served primarily concessionaires and their employees, not the general population. Successive governments were simply unable—due to either lack of political will or lack of ability—to negotiate better terms in the national interest. Therefore a kind of dual economy materialized, and although Tubman's Unification Policy had cracked open the door for indigenous people, they still faced discrimination and power still rested firmly in certain, limited hands.

BY THE late 1960s, the economy was again in a slump; as is always the case, those on the bottom of the social ladder were feeling it first and worst. Expecting a boom in iron ore and rubber prices, the country had borrowed heavily. From 1959 to 1963, Liberia went more than $125 million into debt to finance, among other things, roads and grand new public buildings, including the $6 million Executive Mansion. By 1968, the country's debt had climbed to more than $225 million and President Tubman instituted what he called "austerity measures" after rubber and iron ore prices collapsed.

What's more, the Tubman style of political largesse and patronage was no longer sufficient to keep people from being dissatisfied with the social order and its impact on their lives. Steam was building up, slowly but steadily; the lid on dissent was trembling and threatening to blow.

That Tubman was beginning to feel vulnerable was evident. There had been one major assassination attempt against him in 1955, which set the stage for the massive security network he later implemented. There was another serious military crisis in 1963, in which several leading officials in his administration were implicated. Then, in 1968, came the arrest and trial of Henry Boima Fahnbulleh, Sr.

Fahnbulleh was a distinguished diplomat who had served as ambassador to Sierra Leone and was serving as ambassador to Kenya and Tanzania when he was arrested and charged with leading a tribal movement to overthrow the government. During the trial he was accused of covert contact with Communist countries, and books found in his house, such as *The Decline and Fall of the Roman Empire,* were introduced as evidence. He was convicted and sentenced to twenty years' hard labor, and his property was seized.

Many viewed the trial as an attempt by Tubman to placate the old-line power brokers who were angry at his opening the doors to native people; the trial, it was thought, was a signal to Fahnbulleh, and other native strivers, to get firmly back into their place. Students who attended the trial were so surprisingly vocal in their support of Fahnbulleh and their criticism of Tubman that they were warned that they stood on the edge of disloyalty.

In 1969, I represented the Treasury Department at a conference organized by the Harvard Institute for International Development. The HIID initiative, having been working in Monrovia for about five years, had been begun by Edward Mason, a Harvard economist who helped found the John F. Kennedy School of Government. Mason believed the university had a duty to help developing nations, so he founded a program through which Harvard faculty provided advice on economic and other development issues.

Among the Harvard people in Monrovia at the time was the economist Gustav Papanek. Papanek, now president of the Boston Institute for Developing Economies and professor of economics emeritus at Boston University, suggested to President Tubman that the administration conduct a conference on the country's economic future. Perhaps it was clear to outsiders, in a way it was not to those in power, that Liberia stood at a dangerous crossroad. At any rate, Papanek suggested the conference, the president agreed, and I and several others were invited to speak.

Recently Professor Papanek told an interviewer that I would not normally have been invited to speak at such a conference, given the relatively junior position I held in the Treasury Department. But, he said, he saw something in me that made him want to encourage my growth.

"In my experience there are a small handful of people in any government who are really smart and understand how an economy functions. There are also a handful who are hardworking and a handful who are thoroughly honest. The number of people who combine these three character [traits] is quite small, but she was clearly one of them."

When I tell the story of my speech today, people inevitably ask, "What were you thinking? Weren't you afraid to stand up and criticize the government like that?"

But at the time I did not think of my actions in terms of courage or of fear. The economy was in trouble; the country was standing on the wrong foot and wobbling. The policies weren't working. Someone needed to stand up and say something, and I saw no reason why that someone should not be me.

I still have a copy of that speech, along with most of the major speeches of my long career. I suppose I felt they would be interesting to look back upon one day, if I survived, and they are. This one runs a mere three pages, typed with the little manual Underwood I used back then.

I began by explaining that, as in other undeveloped countries, success in economic development in Liberia depended upon a partnership between the government and its people. The people's role was to invest in learning and trying out new crops and techniques and undertaking efforts and investments that would increase production. The government's role was to levy taxes to provide infrastructure, financing investment in agriculture, housing, education, and manufacturing.

But for such a partnership to succeed, I said, people must believe their government to be honest, efficient, and willing and able to mobilize the resources it collected for the common good. "I think the deliberation of this conference, if it is to be meaningful, while endorsing the achievements in economic development in recent years, must correspondingly recognize the failure which may have led to relative economic stagnation."

At the time revenue growth had declined to about 6 percent annually, from a high of nearly 21 percent per year during the 1950s and early 1960s. Beginning in 1967 or so, the Tubman government had begun taking steps to combat the decline, including increasing taxes on foreigners and foreign-owned corporations and introducing penalties for late payments. But no one had asked why the decline had occurred in the first place. Perhaps it was the collapse of the world markets for iron ore and other exports. Perhaps it was a ripple effect of the government's long-standing tax loopholes for foreign businesses. Or perhaps, I suggested, it had something to do with government corruption and our system of kleptocracy. There was only so much one could steal from an economy and expect the economy to keep prospering.

"With little regard for this situation," I continued, "we still indulge in obvious public wastes such as innumerable foreign embassies, a constantly swelling civil service and other abuses of meager public funds such as payroll padding and outright stealing of public monies."

Many years later, Professor Papanek told me he recalled me putting it this way: "Liberia will never develop until our government stops being a kleptocracy." Those weren't my exact words, but that was certainly the gist. Professor Papanek said it was the first time he had ever heard the word, and he certainly had never expected to hear it from a junior female deputy in an African government.

Well. The ripples through the conference hall were palpable. But I finished my speech and sat down, and the conference continued. Nothing happened immediately.

Afterward Professor Papanek found me in the hallway. Taking my arm, he pulled me off to the side, away from the milling crowd.

"Wonderful speech," he said.

"Thank you."

"It does, however, leave me with some concern about your immediate future," he said tactfully. "I wonder if it would not be a good idea to leave Liberia for a while. As soon as possible."

His concern touched me, though I was—perhaps naively—far less worried for my safety than he. The risk was there, I knew that much. But I was, after all, a woman, and our society did not jail women! Not for merely speaking out!

True, President Tubman had used his security forces to intimidate some political opponents and drum up "evidence" for the treason prosecutions of several others, including Ambassador Henry Fahnbulleh. But this was different. I was only trying to avert future problems. At the most, I thought I might be called before the powers that be and admonished.

Nonetheless, Professor Papanek was insistent. And he said he would look into getting me a fellowship to Harvard, where I could study economics more formally and advance my ability to effect change back at home.

"If I can arrange it, will you go?" he asked.

Would I go? Of course I would! I leaped at the chance to continue my education, and especially at one of the most prestigious universities in the world. What an opportunity!

Professor Papanek said he would try to gain my admission to the Edward S. Mason Fellows, Harvard's oldest and largest international program. This excellent program offers a year of study and the chance to earn a master's degree in public administration to public service professionals from developing countries around the globe.

But first I needed to take the U.S. Agency for International Development scholarship exam, which I did and upon which I scored the highest marks recorded to that time. Then I needed to shore up my undergraduate credentials. I enrolled at the Economics Institute in Boulder, Colorado, where I spent the summer preparing for graduate training. I enjoyed Boulder; it is a beautiful city with huge, awe-inspiring mountains. On a rare break from my intensive studying, I decided one day to try climbing one of the mountain paths. The high, thin air made me so dizzy and sick that I nearly passed out and had to be helped down. I never climbed a mountain again.

We decided it was best that Rob go live with American friends in Elkton, South Dakota, where he could go to high school and stay out of trouble. This is why I sometimes call Rob my American son, because he spent his formative years in that country and is far more American than the rest of us.

After completing my studies in Boulder, it was on to Cambridge. My year at Harvard was a tough but incredible experience, one for which I am grateful and that I will never forget. Not only did I deepen my understanding of economics with challenging courses in subjects such as micro- and macroeconomics, economic development, and quantitative methods, but I broadened my overall knowledge of government and governing with courses in leadership, strategic management, public policy, and data analysis.

I discovered the work of J. Gus Liebenow, considered then and now the dean of Liberian studies. In 1960, Liebenow had carried out research in Liberia and Sierra Leone on the development of political leadership. He had become so intrigued, not to say fascinated, by the complex dynamics of the Liberian political system than he went on to publish several articles about it. His 1969 book, *Liberia: The Evolution of Privilege*, was considered the seminal work of its time on the subject.

Professor Liebenow's central thesis was received with some controversy at the time in the outside world. But his words resonated with me because I knew full well the truth of what he theorized: that a small, elite settler class maintained an essentially colonial relationship to the great majority of indigenous Liberians, monopolizing power and privilege even in the face of superficial societal change.

Liebenow also put into stark relief the truth about the Liberian political system, the lifeblood of our land. As Liebenow pointed out, having a political base was not simply an advantage in Liberian society but *the* necessary foundation for success. "An individual enjoys a position of influence in church, in business, or in a fraternal society because he holds or has held office in the Liberian government or in the True Whig Party. Prominence in a non-political structure is reward for excellent performance in the political arena. The monopoly of the political leadership over religious, economic and other institutions guarantees that those structures will not fall under the sway of those not committed to the perpetuation of the existing distribution of political power. The independent non-political entrepreneur simply does not survive. He is either crushed by a series of legal or illegal barriers or he gets into line."

One year later, in a critical speech that would get me into serious trouble and again alter the course of my life, I would quote extensively from this important work.

I had arrived at Harvard at an exciting time—it was the first year of the Afro-American Studies Department, created after black Harvard students banded together to agitate for change. Among the first professors to teach in the new department was Dr. Martin Kilson, a professor of government who was also the first African American granted full tenure at the college in 1968.

Dr. Kilson helped me realize that to understand a nation's future, one had to first understand its past. I plunged into the study not only of Liberian history but of the history of West Africa. It is no great exaggeration to say that I learned more about my country in the book stacks at Harvard than I had in my entire life back home in Monrovia. It is no small irony that I had to cross the Atlantic to America to learn about my home.

My high school, the College of West Africa, was an excellent school but certainly an elitist one; in truth, all formal education in Liberia at the time was based upon the elitist settlers' version of culture and history. The subjects we studied, the lessons we took, the books we read were either American—little houses in the woods, snow-covered mountains, and other unfathomable things—or Americo-Liberian.

History education in Liberia was largely shaped by an African-American educator named Doris Banks Henries. Henries, born and educated in the United States, came to Liberia as a Methodist missionary and met and married Richard Abrom Henries, a Liberian who became a long-tenured Speaker of the House of Representatives and a major political force. Mrs. Henries spent the bulk of her remaining life in Africa, pioneered the collection of Liberian poetry and folktales, and penned dozens of essays, poetry, histories, and biographies, including President Tubman's official biography. But her greatest effort was the shaping of Liberian education, in particular history.

In more than twenty-seven educational books, Henries crafted the history of Liberia, in which the country sprang forth from the unnamed, unexplored, uncivilized forests of West Africa in 1822. This Liberia fought to survive through terrible attacks by the natives, including the legendary battle of Fort Hill, which took place on December 1 of that year. In that story, the courageous settlers were outnumbered and on the verge of being overwhelmed by the barbaric Africans until a woman named Matilda Newport fired a cannon with her pipe. Victory against tremendous odds was the lesson of that tale. (Over the years both the meaning and the very veracity of this tale were increasingly and passionately questioned, leading eventually to an end of the national celebration of Matilda Newport Day.) The other side—the fears, frustrations, and resentments of the local people, their history and culture—was not taught.

But at Harvard I discovered the true richness and diversity not only of Liberia but of all of West Africa. I learned about the great empires of Songhay, Ghana, and Mali, which flourished in West Africa from the fifth to the sixteenth centuries and encompassed land that borders on present-day Liberia. I learned about the breakup of the Mali empire, and how first that breakup, and later the arrival of European traders—especially slave traders—interrupted the development and political evolution of West Africa and scattered its people into small, autonomous ethnic communities that were far more easily dominated than an empire would have been.

I learned that the history of settler-African relations in Liberia was vastly more complex and more shaded than the Christianity-over-paganism paradigm we had been taught. I learned that the early chiefs with whom the settlers had bargained for land—"King Peter," "King Bango," "King Tom," and other kings whose real names are lost to history—did not realize that in trading their lands for guns and trinkets they were trading those lands for good. They believed the settlers were coming to work with them, not to strip them of their land and authority. When the situation proved to be otherwise, some chiefs saw the future and helped the settlers, teaching them to survive in the new environment. Others opposed those they saw as invaders, fighting intensely and resisting the settlers' domination as long as they could.

All of this was missing in the history I had learned as a girl in Monrovia. But at Harvard I was able to go into the stacks and look at some of the early writings by Americans and Europeans who had traveled into the West African region before the settlers arrived and tried to describe the empires they found.

In essence I came, finally and fully, to realize that Liberia was not just a settler nation. Living in America, watching from the sidelines as black Americans struggled to create and establish their own, empowering identity, I saw how critical it was that my own people know their true history.

Liberians are Africans. Those who were there before the settlers came were Africans. And the settlers themselves, though they brought in different values and different cultural understandings, they too were, in crucial ways, African. It was the reason they came home, and, had they embraced that heritage, history might well have recorded a different tale. Instead they tried to dissociate themselves from all that was part of their blood.

One can understand where they were coming from in this regard: to them, that heritage connoted slavery, and slavery connoted subservience. They were running away from that. But in the process they also ran away from part of themselves.

All of this I learned at Harvard, and it clarified for me, more than ever, our nation's history—and our future unless we changed things and quickly. I wrote a paper for Dr. Kilson both reflecting and solidifying my thinking on all that I had learned and absorbed about my beloved homeland. Liberia had been a golden opportunity, a country with a great endowment in both its people and its land and with the possibility of becoming a beacon to the world. And we had squandered it. We had had so much and had done so little.

The paper also discussed the divide in our nation and made the point—increasingly clear to me from the radicalized campus at Harvard University—that the tensions roiling America and much of Africa would not sidestep Liberia, wishing aside. Change was very nearly upon us. The advantages of the settler elite class were beginning to crumble. Indigenous children had been exposed to education and opportunity, and they were now asserting themselves. The turning point was close at hand.

I did not have much time for a social life that year at Harvard; I spent most of my time in the stacks. Outside the classroom life was little more than studying or attending the numerous lectures and seminars by eminent persons from all around the world, such as John Kenneth Galbraith. On the weekends and holidays I would join the small Liberian community living in Boston for church or for some outdoor activity. I was close to the family of Olubanke King-Akerele, who was studying at Brandeis, and we all spent time together during my Harvard year. Olubanke was the granddaughter of Liberian President Charles B. King, who was charged with sending Liberians essentially into slavery during the 1930 scandal that had rocked the nation, and she and I spent many hours addressing the unacceptable policies and practices of the past. Later Olubanke would work with me at UNDP, and today she is my minister of foreign affairs. I also serve as godmother to her oldest daughter, Bahia. Olubanke continues to work intensely hard, as if making amends for the sins of her grandfather.

At the end of the year I was an A student in every course I was taking, save one: econometrics. According to *The Economist's Dictionary of Economics*, econometrics is "the setting up of mathematical models describing economic relationships (such as that the quantity demanded of a good is dependent positively on income and negatively on price), testing the validity of such hypotheses and estimating the parameters in order to obtain a measure of the strengths of the influences of the different independent variables."

In other words, econometrics deals with the quantitative part of economics, the part in which one does formulas and graphs and curves and all those types of things—the part which my education had not really prepared me for, despite my time spent in Boulder. I worked as hard as I possibly could; still, by the end of the semester I had no idea what grade I might receive. Finally I telephoned the professor and said, "Look, I am an A student in every other course. If you fail me in econometrics, you know you're going to ruin my transcript!" He laughed off my brashness and, in the end, gave me a B. I'm not sure I deserved a B, but I was happy to have it. The rest of my grades were As, and so my record at Harvard was quite good.

My sister was visiting the United States just as the year ended, so we decided to sail home to Liberia together. The first day at sea we stood together on the deck, watching America recede on the horizon. It was July and the air was hazy and warm. I was smoking a cigarette at the time, having become a pack-a-day smoker while burning the midnight oil at Harvard. My sister was appalled at my habit.

"Smoking is terrible for you," she would say over and over again. Although the dangers of cigarette smoke were not as well documented then as they are now, Jennie knew them. She was a nurse and my unofficial health care adviser. "You need to stop."

"Okay," I said to her. "This cigarette I have will be the last cigarette I ever smoke in my life." I finished the cigarette, enjoying its taste. Then I put it out and never smoked another one.

It was on that great ship sailing toward home that my sister and I received word that a major event was at that moment shaking Liberia. It was July 23, 1971. We were having dinner at the time and noticed that one of the stewards kept first looking at and then beckoning to us. Finally he came over to our table and bent low to whisper to us both.

"I'm sorry," he said, "your president has died in London."

My sister and I put down our forks in shock.

"Is this possible?" asked Jennie. "Can this be true?"

We could, neither of us, quite believe it. Neither could the common folks of Liberia, who had known President Tubman all their lives as invincible. They concluded that the radio that had announced his death was lying and refused, for a time, to believe it.

Tubman was seventy-five years of age when he died—not young, perhaps, but far from ancient. Apparently he and his wife had traveled to London so that he could be treated for prostate cancer, but, of course, we knew nothing of that. He died from complications during the surgery.

President Tubman had presided over our country for twenty-seven years, years of patronage and oppression, of Old World charm and ironfisted control. He was a president deeply beloved by the people, but in truth, he was all pageantry and pomp—a characteristic that still lingers with us today. He spoke well of unification, of ending the cleavage that had so long divided the country, but did little in the way of concrete actions toward those lofty goals. Tubman had led what was essentially a benevolent dictatorship during the days when Liberia's resources were plentiful and the prices for its raw materials high. He had possessed the power, the money, and the capability to move our country forward along the path of real political and economic unity but had failed to do so. For years and years the only paved road into the interior not going to a rubber plantation was built by Firestone and stopped at Tubman's rubber farm. Our exotic wildlife, including the pygmy hippo, could be viewed not in a national zoo but only in Tubman's private zoo on his farm.

In short, President Tubman could have done a great deal to develop Liberia for all its people, but he did not.

Still, twenty-seven years was a long time for me and Jennie—our entire adult lives and no small part of our youth. As we sat together in that dining room that evening, we found it difficult to imagine Liberia without the man some called the Maker of Modern Liberia. Difficult but exciting, stomach-knotting, exhilarating, and most of all strange. Utterly, utterly strange.

We both knew anything could happen, for good or for ill. All across Africa—indeed, all around the world—the times were changing. Young people were standing up and demanding to be heard. Peoples who had been disadvantaged and denied were asserting their rights, claiming their own place in society, reclaiming their ownership and entitlement. It was a wild, unsettled, turbulent time, and now Liberia was thrust into the middle of it. One thing was certain: we were sailing home to a whole new world.

Jennie and I sat together in that dining room, praying for the soul of our departed president and praying even harder for our families and our land. We were anxious but not frightened, not really. Like most Liberians, I suppose, we felt in some way shielded from the worst manifestations of evolutionary struggle and change. We always felt that if anything really terrible began to happen, if ever things went seriously awry, America would come to our aid. America was our great father, our patron saint. It would never let us suffer.

That's what so many of us in Liberia thought.

But then we found out that everyone has to stand on his own.

# CHAPTER 4

# THE TOLBERT YEARS

I T WAS a new day in Liberia, and Jennie and I arrived home eager to be part of it. Waiting for me when I returned was the offer of a new high-level position in the newly christened Ministry of Finance (formerly the Treasury Department). I would become deputy minister, charged primarily with responsibility for fiscal and banking policies. My boss, the minister of finance, would be Stephen Tolbert, the proud son of an old settler family, a University of Michigan graduate with both bachelor's and master's degrees in forestry and cofounder of a fishing venture that had grown into the first Liberian-owned multimillion-dollar conglomerate.

According to a flattering 1971 story in the *New York Times*, Stephen Tolbert launched his business career in 1955 when, as deputy minister of agriculture and commerce, he asked President Tubman to commit government funds to study the possibility of building a commercial fishing industry in Liberia but was refused. In response, Tolbert resigned his government post, borrowed $80,000 from a Swiss bank and another $5,000 from his brother, and launched into business with a chartered fishing boat. From that beginning grew the Mesurado Group of Companies, which, by the time Tolbert reentered government as the minister of finance, held interests that included not only fishing but frozen food, detergent, animal feed, and commercial agriculture.

"I knew nothing of business, but I felt obliged to see something done about the project," Tolbert said in the article. "All I had was an idea and some guts."

The brother from whom Stephen Tolbert had borrowed $5,000 in 1955 became, in 1971, our new president.

William R. Tolbert was an ordained minister and former president of the Baptist World Alliance, a quiet, religious man who had spent twenty years in Tubman's shadow as vice president. No one knew quite what to expect when he took the head office in the wake of Tubman's death.

That the Tolbert family was part of the elite settler class was undisputable, though, like many Liberians, the family also had some native roots. What was not clear was how William Tolbert himself would choose to steer the country through the disconnection between the two groups.

Right from the start Tolbert sought to announce that a new day had dawned. At his swearing-in ceremony he startled the guests by wearing a short-sleeved, open-necked safari suit, in one fell swoop abandoning the formal Western attire of top hats and tailcoats that Tubman had preferred and signaling his identification as African.

Even before the inauguration he went to West Point, then the worst slum in Monrovia, and spent an evening visiting the people who lived there. "I want you to know I identify myself with you," he said, according to a story in the *New York Times*. "If you are poor, I identify myself with your poverty, and together we should work to better our conditions." He gave $50 each to several families and urged them to use the funds as seed money for some self-help effort.

On a more substantial level, Tolbert quickly did away with the requirement that civil servants contribute a month's salary each year to the True Whig Party. He eliminated the PRO program, forbade government employees to use official vehicles for private use, and strictly insisted that public employees—contrary to long-established practice—actually show up to work on time. These were in some ways symbolic changes, but they sent the message that the old way of doing things would no longer be acceptable. Reform was on the way.

"Speedy" Tolbert, as he came to be called, declared "war on ignorance, poverty and disease." He called for contributions to "Rally Time," a development campaign, and announced several policies designed to improve the living conditions of the vast majority of Liberians, the upriver and interior rural people who formed the nation's core. His "Mat to Mattresses" plan was meant to lift the poor from the sleeping mats in their huts and shacks onto mattresses in solidly constructed, low-income public housing estates, some of which were built in the early years of his administration. His "Total Involvement for Higher Heights" plan was intended to involve all Liberians in the fruits of economic development. He called for the building of new public schools and clinics, the growth of Liberian businesses, and the flowering of opposition political parties.

Tolbert also promised to wean Liberia from its economic dependence on foreign-owned companies and foreign imports, and to make the country self-sufficient in food production and other basic necessities.

All in all, it was a bold plan. As a man and as a president, Tolbert was a far more progressive-minded individual than Tubman had ever been. His was a true development agenda; he tried to the best of his ability to create a government of true reform. Given a different set of circumstances, he might have succeeded. But even one short year into his administration it was becoming clear to those of us paying attention that many of his plans would be hijacked by the system, never to materialize.

Any person stepping into the job of president in 1971 would have needed tremendous courage and strength to break the cycle of corruption, nepotism, and privilege that had so long held the country in a stranglehold, while at the same time navigating the waves of change coming his way. Liberian society was more unsettled than it had ever been. What had been a largely docile, uneducated population of young natives had now been radicalized, bringing to memory all the cleavage and disadvantages of the past and using that as a political tool. It was a highly combustible time.

Tolbert had spent many years as Tubman's right-hand man and

despite himself was connected to those who had profited greatly from the old way of doing things. Every step he took toward true integration of the country was vigorously and tenaciously opposed by the ruling oligarchy. These people were members not only of his inner circle but of his very family: one brother who was minister of finance and one of the nation's top businessmen, another brother who was president pro tem of the Liberian Senate, a cousin who had served as assistant secretary of the Treasury, and on and on.

All this made it very difficult for Tolbert to pursue his progressive ideas. Later, when the terrible descent began, his brother Frank would accuse him of "opening the gates of Hell" by consorting with natives. This was Tolbert's contradiction: he was a man caught in the middle, willing but unable to move fully forward, unwilling to retreat.

In November 1972, I was invited by my alma mater, the College of West Africa, to give the graduation address. Most graduation speeches are an opportunity to look back fondly on one's days at a school, to commend the students for their achievement and extol their academic excellence before urging them to go out and do some vague, undefined good in the world. But I was not interested in that. I was working hard at the Finance Ministry, deeply concerned about where the country was headed, and in no small part frustrated at what seemed to me a lack of urgency on the part of the powers in charge about changing things. I considered my address an opportunity to challenge the government in a way it could not ignore.

The title of my speech was "Some Fundamental Constraints to a Liberian Society of Involved Individuals." I am going to quote somewhat liberally from it here because it was, in the course of my life, an important turning point, one at which I set my feet upon a path from which there was no turning back. I was climbing the ladder of success fairly quickly at that stage. The thing to do, if I wanted to be successful according to the terms of success that had long dominated our society, was to go with the flow. But I could not do that.

In the speech, I began by thanking the school's board of trustees for inviting me, knowing that I was the type of person to always speak my mind. Even then I realized that their invitation indicated that progressive thinking and desire for change were affecting even some of us who had been fortunate enough to step into the privileged class. I was not alone.

I told the students that the nation stood at the verge of a crisis, a crisis over which we might have no control. Tensions were mounting—economically, socially, and politically. Yet within this tension existed the potential for change, but only if we, as a nation, were able and willing to push aside the curtain of fear and deceit in discussion of our national plight. It was time, I said, to stand up and speak the truth about who we were as a country.

Like the United States, Liberia had been born in a quest for human freedom that prompted difficult and dangerous voyages—in different directions across the same ocean. Likewise, as with those of the United States, our own founding principles had been flawed. Both nations had been founded by men of their time, representing an elite class. They had not extended the liberties they sought for themselves to indigenous people, to all ethnic groups, and to women.

The very symbols of Liberia—the seal, the flag, the motto—are often seen as symbols of exclusion and division. This was not a history that the students or I could help. But what was sad was that 125 years later we seemed unable or unwilling to change what was slowly pulling us apart.

"We have been told that these instruments are too historically sacrosanct to be tampered with. But I ask each of you students in this audience to take a few moments out this evening to read the constitution of your country. Then compare the sacred rights embodied therein with the practices with which you and I are so familiar. Examine how these supposedly sacred rights have been ruthlessly prostrated. Perhaps you, like I, may conclude that those who now make empty, sanctimonious claims about rights are worthy of only our deep contempt."

What I wanted to do was to bring some attention to what I saw as a mounting crisis. President Tubman had opened the door to social alteration decades before, and President Tolbert had pushed it open even more—but the changes were superficial at best and the pace was far too slow. Everyone knew this in private, but no one dared say so publicly. I would.

Only days before, I had attended the University of Liberia's graduation ceremony and witnessed a moving moment when the university choir had sung one of its songs in a local language. Everyone in the audience had applauded with great warmth and pride. But lost on no one, at least not upon me, was the fact that it was the study of French that was stressed in the university curriculum, not Kpelle, Vai, or Bassa. One song sung at graduation was window dressing at best.

Every country needed some great cause to bind its citizens, I told the graduates. For most African countries this great cause was the struggle to throw off the yoke of colonialism. Liberians lacked this rallying cry and so would have to search and adopt other approaches to a strategy for true national integration.

When we could learn to avoid, in the subconscious, making distinctions between one another on the basis of last or first name, when all of them graduating that day could compete for higher education scholarships and be assured that the yardsticks for measuring who would receive such a reward and who would not would be based solely on merit, when every youth of the land was exposed to equal opportunities in school, housing, and employment, when we learned to respect character over family name and moral achievements over material accumulation, then, and only then, would the walls of social constraints come tumbling down.

Finally, I thought it important to give the graduates a perspective on the history of the nation's economy. Many people thought we had been, and would continue to be, saved by Firestone. But forty-six years after the entry of Firestone, the Liberian economy was still characterized by large, enclave primary industries, by foreign traders operating on a caveat emptor basis, and by a few wealthy Liberian entrepreneurs. The income disparity between the haves and the have-nots was huge and dangerous.

I urged the graduates to reject materialism not only for their own sakes but for the sake of their nation. I warned, prophetically, that the widening of economic differences would create unbearable tensions. My generation might not be on the scene when it happened, I said, but God help those who were.

That was in 1972. As it turned out, the violent and horrific coup d'état would take place only eight years later, with my generation very much still in play.

All the time I stood at the podium making this speech, I was aware that behind me on the stage sat some of the most prominent, conservative people in our society, people who were very close to the old order and who zealously guarded their privileges. Their disapproval was palpable; it rolled toward my back in waves. Before me sat the students, who were clearly excited by what I was saying but did not dare show it. They were facing the group behind me, and they could see the growing storm. When I finished, the students all clapped enthusiastically. Behind me, the silence was deafening.

It was a tough speech, but it had to be. I intended my words to be a kind of warning, a set of flashing lights meant to slow down the speeding train.

Instead, they turned out to be a prediction. A sad, terribly accurate one.

THE NATION's reckoning was still years away at that point. My own, however, was close at hand.

Having been down this road before, I knew before giving the speech that I would get into trouble afterward. Still, I think I had no real understanding of the extent of the punishment that could have been imposed on me, the extent of reprimand that some members of the Tolbert government wanted to inflict.

Before the day was out, the speech had been banned. But one of the graduating students, Victor Jesus Weeks, son of the minister of foreign affairs, got a copy of it and printed it in a small newspaper the next day under the title "Print That Speech!"

That morning I rose, dressed, and went into work as usual, not knowing what to expect. No sooner had I sat down at my desk than the summons arrived.

"The minister would like to see you in his office," the secretary said. "Right away."

When I entered Tolbert's office, he was clearly furious. "Go and bring me that speech," he demanded. After I complied he read some of it aloud, clearly for effect because he had obviously already read it. After a few lines he stopped and began to castigate me.

"You know, people do not usually challenge the government they're in!" he thundered. "You don't say things that will put the public against the government. We are not going to tolerate that. This is very serious."

I was sent back to my office, but the matter was not over. Very soon the issue of my speech was taken up in a high-level meeting of the president and his top ministers. Some members of the cabinet were furious. The word "saboteur" was hurled around; I was accused of everything from nerve to effrontery to treason of the highest kind. Some called angrily for my head, demanding that I be immediately and summarily fired from my post. Others suggested that tossing me into prison might be a more fitting response.

"We cannot allow this one person, this woman, no less, who is herself in the government and who is benefiting from the government, to stand up and criticize the government," they said. "In effect, she's criticizing herself and everything we all stand for."

But President Tolbert, to his credit, had not merely surrounded himself with sycophants. The government was stuffed with members of the old guard, intractable and increasingly worried about losing their grip. But there were also younger, more sophisticated, slightly more progressive voices. These people had been abroad and received, both formally and informally, the education that such experience provides. Some may simply have understood that the old tactics would no longer work. Others, no doubt, actually believed the things I said and secretly agreed with me. It was those people who provided the cooler heads and softer touch in the meetings on my case. They persuaded the president and the hard-liners in the cabinet that it was not in their best interest to make an example of me. The best thing, they said, would be to leave me alone.

The minister of foreign affairs also happened to be the father of young Victor Weeks, who had published my speech. According to what I was told, after the cabinet members had finished berating me, they began berating young Weeks as well. Finally, Stephen Tolbert asked, "All right, now what do we do?"

Rocheforte L. Weeks took the floor and, speaking of his son, said, "We brought up our children to respect authority, and he should not have done what he did. He needs to be punished, to be taught a lesson. If that means jailing, he needs to be jailed."

Later Steve Tolbert told me that as Weeks said those words, he changed his mind instantly about the need for us to be punished. He stood and said to his brother, "Mr. President, let me say to Minister Weeks that if that were my son, I would be proud of him."

That ended it. The cabinet was dismissed. I was not fired, and neither young Weeks nor myself was even seriously threatened with jail. That's how powerful Steve Tolbert was at the height of his authority. What's more, Steve Tolbert, for all his failings, respected people of strength, because he was himself a man of strength. He respected the positions I had taken in the ministry, even when they conflicted with his own.

In an amusing side note, Steve Tolbert would later decide to sell copies of that speech for $1 each to raise money for a controversial pet development project of his brother's called Rally Time. In one of my life's unexpected moments of saving grace, Rally Time would later help save my life.

Though I held on to my job, I was, nonetheless, deliberately side-lined. At meetings I found my contributions ignored or pushed aside. At my desk I found major responsibilities removed from my care bit by bit. I was being isolated and, eventually, it was hoped, would be squeezed out.

So obvious was the fact of my disfavor that other critics, people working outside the government and beginning to lobby for change, began visiting me and talking about the issues I had raised. These people would never have approached me had they believed me loyal to the old guard and the old ways, but now they came. Together we would sit on the porch of my house and discuss both the present and the future of our land. These meetings, open and public, distanced me further from the ministry.

Among the people I met with was Albert Porte. Porte was a remarkable person, a schoolteacher, journalist, and one-man crusading force who had been advocating for social justice in Liberia since the 1920s. Porte routinely self-published and distributed pamphlets challenging the entrenched privilege of the settler elite class and the oligarchy of the True Whig Party. He had clashed with President Tubman so many times, it was said he never left home without his toothbrush and toothpaste because he never knew when he would be picked up and tossed into prison for disseminating his critical views. At the time of our meetings, Porte was turning his eye on Stephen Tolbert, accusing him of misusing his public office to advance his extensive and exceedingly profitable business interests. In 1974, Porte would publish his famous work, "Liberianization or Gobbling Business?," a pamphlet that launched a direct attack on Stephen Tolbert. Tolbert responded by suing Porte for libel in a massive civil damages suit. The judge assigned to the case was Chief Justice James A. A. Pierre, Stephen Tolbert's father-in law. Not surprisingly, Tolbert won. However, the public's disgust at the verdict was one of the first real steps in the radicalization of the society, a radicalization for which William Tolbert would later come to pay a heavy price.

But these events would all take place later; at the time I was meeting with Porte and others the lid was still on the pot, though rattling.

As the months passed and it became more and more clear that I was being pushed out, I decided, at last, to act. In my position as deputy finance minister I had met several officials from the World Bank. These officials, my colleagues and friends, had come to know me and respect my work. Finally, in the spring of 1973, I telephoned a few of them. "I think I'm in trouble here," I said.

With their help, I applied for, was offered, and accepted a position as a loan officer, becoming part of the first group of Africans recruited by the bank. A few days after announcing my intended resignation from the Finance Ministry, I was at home one morning preparing to go to work when Steve Tolbert stopped by. It was a startling moment, made even more startling by the fact that he did not wait to be received but came right into the house. I was in the bathroom at the time, brushing my hair and preparing for the day. Someone must have announced to me that the minister of finance had arrived, and then, before I had time to react, I heard his voice.

"I came to see you," he said.

"Let me finish, and I'll be right out," I called through the door.

But Tolbert was a forceful man, accustomed to having things done in his way and on his time. He simply pushed the door open and strode into the room where I stood, amazed.

"What is this I hear about you leaving the ministry?" he demanded.

When I told him I had accepted a position at the World Bank—which he already knew but wanted to hear from me—he accused me of having improperly used my contacts to secure the position.

"I could easily stop you by telling them you have misused your official position for personal gain," he said. "But I won't do that. If you want to go, I'll let you go."

I said that I did, in fact, wish to leave, and that was that. In truth, I think that at that point both Tolberts were happy to see me leave both the government and Liberia. They believed that with my leaving went one major problem for the government. In the end, Minister Tolbert actually threw a large reception in my honor, a rousing party to send me off in style.

The World Bank, along with the International Monetary Fund, was established in Bretton Woods, New Hampshire, in July 1944, even as World War II was grinding on. The bank, formally known at the time as the International Bank for Reconstruction and Development (IBRD), was created essentially to rebuild Europe after the devastation of war. Its first loan, $250 million in 1947, was to France so the heavily damaged nation could purchase, among other goods, ships, trucks, radios, coal-mining equipment, copper, tin, synthetic rubber, animal fats, and chemicals.

Liberia became a member in 1962 and two years later received its first loan. The funds were designated to go to road projects, including construction of the road from Monrovia to Roberts Field, the location of our international airport.

Today the World Bank is not a bank per se but a collection of five distinct development institutions, including the IBRD and the International Development Association, which focuses on the poorest countries and provides "soft loans" at low or even zero interest rates. The bank is owned by 185 member countries, all of which are shareholders and who are represented by a board of governors. This board, which makes policy for the bank, generally consists of the ministers of finance or development of the member countries. In the case of Liberia in 1972, this was Stephen Tolbert. But as deputy finance minister, I naturally attended many meetings and worked closely with bank officials on bank projects in Liberia.

I considered my posting to the World Bank to be a period of growth, connection, and respite. I left Monrovia and headed to Washington, thinking to myself "I will use this opportunity to better myself professionally." It was a strategic retreat, not a surrender. I always knew I was going to come back.

With my mother and my son Rob, I moved to Alexandria, Virginia, and set up house in a lovely home. All three of us wanted to live somewhere that reminded us of home a little, with trees and greenery and wide, open spaces, some place less urban than D.C. itself.

Washington itself I found fascinating; it was, and remains, of course, a city of many contrasts—sparkling monuments and decaying neighborhoods, power and poverty, the most powerful of powerful just blocks from the completely powerless. Among the World Bank staff there circulated the usual warnings about avoiding certain areas of town. I never let these bother me or stop me from looking around.

For the most part, however, I was too busy traveling on World Bank business to have much time to sightsee in America.

My first posting was to the Latin American and Caribbean division, where I was stationed as a loan officer to Barbados. The position gave me my first chance to travel to the Caribbean, which I found to be exciting.

After some time I was promoted and assigned as a loan officer to Brazil. In Brazil I was fascinated to learn about macumba, the Brazilian religion that grew from a syncretism of traditional African religions, Brazilian spiritualism, and Catholicism. Of the many forms of macumba in Brazil, the most important are Candomblé and Umbanda. Observing a macumba ritual, I was interested to see how closely related it was to familiar African spiritual worship. There were the outdoor ceremonial site, the offering of flowers and other gifts to spirits, and, most of all, the dancing. Throughout both the Caribbean and Brazil I often encountered powerful and familiar glimpses of home. The African diaspora indeed stretches far and wide.

After some time working in Brazil I was again promoted, this time to senior loan officer in the East African division. For me this was almost like a coming home—as well as a valuable chance to begin building strong professional relationships—in Kenya, Tanzania, Uganda, and other countries—that would last me the rest of my life.

People—usually women—sometimes ask me if, during my long climb up the career ladder, I ever bumped into any glass ceilings or encountered resistance to my taking a seat at the table because I am a woman and African. My answer is that I am sure there have been those who suspected me of being a token or who resented my having the positions I had. But I was usually too busy to worry about them.

In the case of the World Bank I was one of the first group of African midcareer professionals recruited, and the bank, to its credit, did not simply walk us through the door and leave us idle. The OAU had been pushing for an African presence inside this institution, which was exerting so much financial power over Africa, and the bank responded. It wanted us there; it wanted to be seen as having the diversity an international agency should have. We were all given a great deal of opportunity.

It is true, however, that reforms initiated at the top do not always filter down quickly or easily. One did have to face a lot of people who had been in the bank for a while and who made clear their feelings that these new people coming through the door didn't really know their stuff. The first few months at the bank, we all felt people watching to make sure our professional credentials were what they should have been, watching to see whether we could cope with our assignments. But for many of us, myself included, we were working too hard to pay much attention to any rumblings. I spent many a night after work studying reports and reading an endless stream of documents to deepen my knowledge of international economic policy.

I think it also helped that my initial posting was to Barbados, a small country with a distinct African heritage. That connection, that shared heritage, helped me establish very good relationships with the people I worked with, and the bank valued good relationships with its client countries quite highly.

It took perhaps a year for me to really find my rhythm at the bank, to understand the inner workings and to master the buzzwords and the lingo of the agency. After that the job became increasingly manageable, and as my confidence grew so did my professional credibility and hence the acceptance.

The promotion to Brazil put me at the bottom of another hill to climb, of course. Brazil was a much larger country than Barbados, and my Portuguese was rather poor. One of my first assignments was to oversee one of the bank's first nutrition projects in the country, and there was, among some of my colleagues, concern that I did not have the ability to present the project properly to the board. One colleague suggested quite openly that if I went before the board and got tough questions, I would falter and fall.

But the head of the division, Alan Berg, stood up for me. "Let's give her a break," he said.

Then he pulled me aside. "Listen, I'm taking a big risk with you," he said. "If it doesn't go well, it's going to really cost me some major trouble."

I thanked him and went home to prepare, and in the end the meeting went off without a hitch. In fact, I handled the presentation and the tough questions so well that afterward Alan bought me flowers to celebrate.

There were some meetings in which one felt a little bit inadequate, when some of the larger issues were being discussed and you felt you really should speak up. At those times I sometimes felt the bank was being much too domineering. Too often bank employees went to sovereign countries and spoke to presidents and ministers as if they were lecturing children on what they should and should not do. This was arrogance, plain and simple, and I knew it, but I would sometimes remain quiet and say nothing because I didn't want to get into trouble with the bank.

Many critics of the World Bank complain long and loud against its policies, suggesting even that its goal of reducing world poverty is just a shield for its true goal of promoting U.S. business interests. I say that, to be sure, it is very clear that the World Bank serves the interest of the major Western powers and that its primary decisions are made by them. The very structure of the bank ensures as much: its president is always an American, nominated by the president of the United States. What's more, the United States owns such a large percentage of shares that it can exercise, in effect, veto power over any major decisions taken by the bank.

There are many bank positions some of us, as staffers or as government financial officials, would question over the years, such as the bank's insistence on completely open markets. "No subsidies" for poor countries, the bank says. "Subsidies are bad for open markets." But that tenet does not seem to apply at home in the United States and Europe, where agriculture is still heavily subsidized. So the, shall we say, unfairness is there. No doubt.

My belief, however, is that countries must either take it or not take it—you can't do both. Poor countries are at a disadvantage—they want the funding from these Western institutions so desperately that they are not willing to say no to all the strings that come along with the wonderful gifts. So the leaders take the funds and make the promises and afterward there are reversals—they cannot or will not go along with the terms of agreement. The funding stops. The development process is interrupted. And the country is worse off than it was before.

This is a capacity issue. Most African countries simply did not have that critical mass of trained people with the experience to take on these international institutions and negotiate head-to-head. Asia was much further ahead in this regard. Asian countries didn't have to take what was handed to them. They determined their own agenda; they set the pace and told the bank what they would and would not do. Africa was behind because we didn't have the capacity to resist. Happily, that's changing. Today, Africa is pulling itself up. Botswana, for example, a small country with a driven economy, does not need the World Bank. It makes us proud.

All in all, however, my time at the World Bank was a very rewarding professional experience, allowing me to further deepen my understanding of international finance and redevelopment and expand my professional portfolio. It was also a lot of fun, traveling around and seeing the world. But at the back of my mind, I was always ready to go home.

In April 1975, Stephen Tolbert and five others were killed when their twin-engine Cessna crashed into the Atlantic Ocean just after takeoff from Greenville. After mourning his brother, President Tolbert appointed James T. Phillips minister of finance. Phillips was considered one of the more reform-minded members of Tolbert's circle. He called me in Washington and asked if I would consider rejoining the Finance Ministry. I wanted very much to go home, but having watched the Tolbert government from afar, I remained unsure of its commitment to real reform and wasn't sure I wanted to cut my ties with the World Bank. After some discussion among all the parties, it was agreed that I would be seconded as a World Bank staff member to work for the Liberian government.

I returned home to a country in flux. President Tolbert's easing of restrictions had emboldened dissenters, which seemed to surprise and even wound him. The libel trial of Albert Porte had galvanized the simmering demand for political reform among university teachers and students, leading to the creation of the first of what would quickly become a slew of civil society organizations, the Citizens of Liberia in Defense of Albert Porte (COLIDAP). When Stephen Tolbert died, the lawsuit against Porte essentially disappeared, but the demand for reform stayed front and center.

Of the many movements that sprang to life during this turbulent period, one of the most active was the Movement for Justice in Africa, or MOJA. MOJA was founded in 1973 by an economics professor at the University of Liberia, Togba Nah Roberts, an ethnic Kru who remains president of the organization to this day. Among other early MOJA leaders were two other prominent Liberians: Amos Sawyer, a political science professor who later became head of one of the transition governments during the civil war, and Henry Boima Fahnbulleh, Jr., the son of the former ambassador charged with treason by President Tubman in 1968. Fahnbulleh, Jr., would go on to serve as Liberia's foreign minister and would run unsuccessfully for president in 1997. Although over the years he would become one of my most voracious critics, he serves today in my administration as my national security advisor.

Both anticolonial and pan-Africanist in its scope, MOJA played a pivotal role in radicalizing the urban and rural poor of Liberia, raising the issues of government corruption, advocating for the nationalization of Liberia's major businesses and demanding an end to government corruption.

Meanwhile, a group of Liberian students in the United States, led by Gabriel Baccus Matthews, created, in 1975, the Progressive Alliance of Liberia (PAL). Baccus Matthews was at the time Liberian vice consul in the United States, but he quit his post after differences with the foreign minister to dedicate himself to political activism, labeling himself a staunch socialist and pan-Africanist and calling for true multiparty democracy in Liberia. As we were both in the United

States, he reached out to me and I participated in meetings of his group. We were sometimes joined by Winston Tubman, William Tubman's nephew, who was then working at the United Nations. I thus became identified with that movement, even though I was not really a member of PAL.

Baccus Matthews would go on to play a crucial role in the tumultuous events to come in Liberia. In a way he was liked and tolerated by President Tolbert, and he used this privilege on many occasions to challenge and sometimes even insult the president, even in the presence of others. The price of rice became his cause célèbre.

The two of us often found ourselves in contention; we did not always see eye to eye, but we both worked long and diligently to serve our native land. But by the time of his death many decades later, I meant it when I called him the Godfather of Liberian Democracy and said he had left an indelible mark on our country's political history. Despite the many contradictions he represented, Gabriel Baccus Matthews will be remembered as one of Liberia's greatest sons.

All this was simmering upon my return to Monrovia, and President Tolbert was struggling in his attempt to respond. To his credit, it was President Tolbert himself who had invited the leaders of PAL to return to their homeland to register as the first opposition party on the books in more than twenty years. I'm not sure he really anticipated what would happen once they took him up on his invitation.

The country was facing severe economic problems. At the same time, President Tolbert was preparing Liberia to host the Organization of African Unity's annual conference.

In some ways Liberia was the perfect place to host such a meeting. The idea for an organization that would bring together the diverse countries of Africa for collective political and economic benefit was first given momentum when President Tubman met with President Sékou Touré of the Republic of Guinea and President Kwame Nkrumah of Ghana in 1959 in Nimba County to discuss the idea. Four years later, in May 1963, the Organization of African Unity was born in Addis Ababa, Ethiopia.

At the inaugural meeting in 1963, Nkrumah called for a union of African states with a strong central government to act in the best interest of the continent as a whole. But the thirty-one heads of state present rejected that idea, establishing instead a loose federation of nations whose sovereignty was guaranteed by a strong—and in some ways primary—policy of noninterference in the internal affairs of other states.

President Tolbert considered hosting the 1979 annual summit to be a great honor and privilege, both for himself and for Liberia. The rationale for his decision to do so, taken in 1975, was said to be twofold. One factor was that Liberia, increasingly marginalized with the coming into independence of more viable African states, needed to reestablish its leadership role in the OAU, as had been achieved by President Tubman during the formative years of the organization. Second, it was argued that domestic economic activities associated with the OAU hosting would provide an impetus for the resurgence of economic growth.

(In truth, the one possible justification for spending all the money and resources to host the OAU summit was that the event was normally followed by the hosting of the African Development Bank's annual meeting a few months later. Unlike OAU meetings, ADB meetings usually bring investors and financiers to the host country, opening up opportunities for investment and development. The ADB annual meeting was scheduled to take place in Liberia the following May, in 1980—but it would never happen.)

Critics dismissed the government's rationale on both counts as not only dubious but unfounded. Nonetheless, President Tolbert pushed forward. Roads were paved, sidewalks built, shops created. An ocean liner was rented as a floating hotel. The grand Unity Conference Center was built outside Monrovia in the seaside suburb of Virginia at a cost of $33 million. The multistory Hotel Africa—which now stands gutted, forlorn, and empty—was built nearby, along with a swimming pool shaped like the continent of Africa and fifty-one beach villas, one for each African head of state expected to attend ($36 million). Billboards touting the various heads of state were constructed along the roads to the airport and conference center.

The cost of the conference proved to be high, albeit much less than the amounts cited in rumors swirling around Monrovia at the time. The final estimate by a major public accounting firm put the price at $101 million. Such a figure would place it right around one third of the annual Liberian budget at the time.

It must be said that at the time few Liberians considered the OAU vitally important. We saw it not as a meaningful institution that could make any difference on the political scene but as a place where a group of leaders—all, of course, men—met periodically to rub shoulders with one another, to showcase their political power, and to socialize. Today, nearly thirty years later, the OAU is changing, becoming far more substantive in its mission of creating a more integrated, vital, peaceful, and prosperous Africa. It even has a new name—the African Union. But in 1979, it was pretty much a place where people went to read long speeches and go to cocktail parties. It was a men's club and, in many cases, a dictators' club.

To me, then, as to many others, it seemed unwise, if not downright dangerous, to be spending so much money hosting this exaggerated men's night out at a time when the Liberian economy was under such serious strain. In a nutshell, we simply did not have the money to host the summit. I found myself taking rather strong positions in opposition, both to the overspending on the summit itself and to the many blatantly inflated or downright fraudulent invoices being submitted to the government by contractors on behalf of OAU projects.

I remember one bill that found its way to my desk. A contractor, British, had written to President Tolbert complaining that the company had not been paid properly. Tolbert sent it to Finance Minister Phillips, who in turn sent it to me for comment.

The bill struck me as outrageous. I decided to use one of the rubber stamps I had bought in a small souvenir shop in the United States that said BULLSHIT. I felt that was an appropriate response to what I was seeing, so I stamped the bill and returned it to the British contractor. Of course the press got hold of the incident; a *Financial Times* story referred to an African executive who used expletives to respond to correspondence. Phillips called me into his office the day the stories appeared and ordered me to bring him the stamp.

"You can't use that in this place!" he said.

So much for the confiscated stamp—but not for the sentiment it represented.

On several occasions I tried to take my concerns to the president himself, who was, again to his credit, an informal and accessible man. During our meetings I would haul out papers and charts and reports, trying to discuss serious economic theory, trying to bring every bit of knowledge and understanding I had about running an economy to bear on our situation. Tolbert would simply smile and reiterate his confidence in me. "Just go and do the best you can do."

Whether this was his response to all his cabinet members or only to me because I was a woman I could not tell, but I sometimes worried that he took the country's economic situation too lightly. Liberia had been politically stable for more than a hundred years, and I believe he counted on that, stood on those more than a hundred years of calm and civility, of docility. It was as if he thought, "Nothing bad can really happen. Our people will just make some noise, that's all." President Tolbert simply refused to see that although the surface was calm, the undertow was strong and getting stronger.

That April, the surface calm broke.

While President Tolbert continued to oversee preparations for the coming OAU summit, Florence Chenoweth, the minister of agriculture, proposed raising the subsidized price of rice from $22 to $27 per 100-pound bag.

Rice is not only a staple of the Liberian diet, it is *the* staple, the foundation upon which most Liberian meals are built. For the affluent there are also chicken and beef, fish and other seafood, and for the poor there may be various local fruits or vegetables or cassava in its many and myriad forms, but for everyone there is rice.

Although as early as 1973, President Tolbert had set self-sufficiency in rice as one of the country's prime economic objects, and although some important steps had been taken toward achieving that goal, by

1979 Liberia was still importing some 25 percent of its consumed rice. Simply put, there was too little money in the crop to encourage farmers to grow much more than was needed for their families. Rice was not subsidized, and the price had been kept artificially low by the government.

In 1975, with a loan from the International Fund for Agricultural Development, the government established AGRIMECO. AGRIMECO was a government-owned, Israeli-run, American-supplied project in Lofa County intended to clear large swaths of farmland and turn it over to farmers, both to increase food production and to decrease the damage to the soil from the more traditional methods of farming. But by 1979, little progress had been made on this score, in part because some of the heavy equipment being used was far more appropriate for use on the wide, flat plains of the United States than the thin, sometimes swampy, sometimes heavily forested soils of Liberia.

By 1979, the Ministry of Agriculture was proposing the price increase as a way of encouraging small and medium-sized farmers to produce more rice. The idea was not only to boost local rice production but to encourage more farmers to stay on the land rather than deserting the villages and farms to seek employment in the cities or on rubber plantations or in iron ore mines.

Critics argued that the ministry's position was flawed on three counts. First, that the increasing trend in rice production did not support the view that the existing price served as a disincentive. Second, that the low supply elasticity for rice meant that there would be little response to the price increase on the part of the producers. Third, that because the number of commercial producers was small (only 10 percent of the total) the benefits would benefit mostly the few large producers (including some in government) rather than the subsistence farmers on whose behalf the ministry justified its position. Mostly, though, critics argued that the increase would greatly harm most Liberians, whose average wage at the time was $80 a month. The announced price increase sent a wave of anger throughout the countryside.

The increasingly vocal radical leaders seized the issue, accusing the government of profiteering on the backs of the poor. Key government officials, including President Tolbert himself, owned large rice farms and thus, intentionally or not, stood to profit from the increase. PAL, led by Baccus Matthews, called for a demonstration to protest the proposed increase. He claimed to be able to import and sell rice for $10 a bag.

In the days leading up to the protest, President Tolbert warned the organization's leaders not to go forward with the plan, saying they had the right to petition but not to demonstrate. Matthews announced his intention to proceed nonetheless.

The rally was called for 3 p.m. on April 14, 1979. But by dawn people had already begun to gather at the PAL headquarters, not far from the Ministry of Finance. I was living at the time in an apartment building right across the street. Because it was a Saturday, I was home at the time. Very early in the morning I noticed the young people gathering in the streets and opened my window to see what was taking place. Something like this had never happened before in Monrovia.

By midmorning the crowd had grown to some two thousand people, mostly young men. At first the crowd was animated but peaceful, chanting slogans and waving signs. But as the crowd swelled, growing larger and larger until, by some estimates, more than 10,000 people lined the streets of Monrovia, the tenor of the demonstration changed. Soldiers in tanks stood by, while the police tried to disperse the crowd with tear gas, to little effect. People began smashing windows, leaping onto traffic lights and pulling them down, looting from the small shops and stores that lined the streets.

From my apartment I watched, shaken and worried for my country. In the weeks leading up to the demonstration, the so-called progressives or radicals had strung a large banner bearing one of the mottos of the radical groups across one of the streets. "Our eyes are open," it read. "The time of the people has come."

As I stood at my window, looking down as the world I knew began to break apart, those words reverberated in my head. "It's coming," I thought. "It's coming."

It was the beginning of the end of the political system that had dominated my country for more than a hundred years. I think I sensed it even then.

More police soon arrived on the scene, supplemented by the president's security forces, which had been given the authority to shoot into the crowd. To my horror, they opened fire on the demonstrators, killing some people instantly. Bedlam ensued, with people screaming and running to escape the flying bullets.

In a newspaper interview the following month, President Tolbert would say that the shooting had been absolutely necessary and resorted to only when a group of demonstrators broke off from the main crowd and headed toward the Executive Mansion.

"When they did not check the demonstration by using tear gas, then the next thought was to fire into the air," President Tolbert would tell the *New York Times*. "That made no effect. Not until they were very near the mansion with whatever plans they had in mind to do, then someone got injured from the security side."

President Tolbert maintained that he had given orders for the police to fire only if they had to in self-defense and then to aim "down in the extremities" to minimize fatalities. But these precautions made no difference. Chaos ensued.

In the end the Rice Riots, as they came to be known, raged all night and into the dawn. Like most Monrovians, I stayed inside, telephoning various family members to make sure they were safe. When at last the looting and violence subsided, at least 41, and possibly as many as 140, people had been killed, with hundreds more injured. More than 160 stores had been ransacked, many of them owned by Lebanese, resulting in some $35 million worth of property damage and theft.

President Tolbert seemed stunned by the violence. The day after, he called the leaders of the demonstration "wicked, evil, and satanic" men who hoped for chaos and disorder so that they might eventually overthrow the government. He ordered the arrest of Gabriel Baccus Matthews and thirty others involved in the protests and had fourteen of them charged with treason. He shut the University of Liberia down for a time, accusing instructors there of using their classrooms as breeding grounds for subversion and revolutionary ideas.

Later, he appointed thirty-one people, including myself, to a commission to investigate the incident and make recommendations that would prevent further disturbances. The commission was headed by an old and eminent citizen, Netesie Brownell.

That June, our commission delivered its report. It was a tough and critical one—the *Washington Post* called it "scathing"  that called for amnesty for the demonstration leaders and investigation of the police director and the ministers of justice, agriculture, defense, and finance. We called on the president to create a code of conduct for all public officials that would begin to tackle the corruption and conflicts of interest so rife throughout government. Quite pointedly, the commission called on President Tolbert very specifically to examine the social impact of the very high visibility of so many of his relatives in monopolistic business ventures.

The commission made a number of specific suggestions for addressing the country's widening economic gulf and the tensions it caused, including ending Rally Time (which people felt had become a system for forced contributions to corrupt government), giving more attention to minimum wage increases, and providing more assistance to small businessmen, especially those whose businesses had been damaged in the riots, and paying its debts to local businesses as soon as possible. We also warned that the completion of all the OAU-related building projects would create a dangerous surge of unemployment that needed to be addressed.

President Tolbert, still reeling, responded to the report by making a major speech to the nation, during which he granted amnesty to Matthews and the others charged with treason. He also lowered the price of rice by $2 a bag and gave pay increases to the police and military.

During the speech, Tolbert tried to address what he saw as the causes of the riots, which he said were socioeconomic in nature. The Liberian people were demanding more schools, more roads, more hospitals, more electricity, and more piped drinking water in the interior, he said, and the government was working hard toward those goals.

But, he said, the people must understand that the government was working "against the cruel realities of economic restraints."

Tolbert seemed to believe the speech had bought him some space and time. Although the commission had strongly recommended that its life be extended so that it could continue making long-term recommendations for addressing structural inequalities, Tolbert dismissed the group, saying its work was done. For all intents and purposes, he put our report aside, ignoring the bulk of the suggestions and recommendations we made. Then he continued preparing for the OAU summit. I believe this was one of his greatest missteps.

After the Rice Riots, Tolbert was forced to adopt new security measures in an attempt to stabilize the situation. What he was unable to do was to show enough concern for the political and economic undercurrents that had brought about the unrest. What he failed to do was to liberalize the government.

The riots proved, once and for all, that the society had been radicalized. There was no turning away from that fact, no putting the genie back into the bottle. Yet the old hard-liners urged President Tolbert—and, according to some rumors, threatened, due to their fear of being overthrown—to become more repressive instead of less, to pull back from the brave but insufficient steps toward a political flowering that he had allowed. It was the worst possible advice.

What is tragic, I believe, is that had Tolbert instead implemented the major report recommendations of the Commission on National Reconstruction, things might well have turned out differently. Had the government seized the opportunity to make major changes, it could have turned the whole thing around and avoided all the bloodshed to come.

THE SIXTEENTH annual summit of the Organization of African Unity opened on July 17, 1979, in Monrovia. Twenty-four heads of state attended, with the king of Morocco pulling out at the last minute. In his opening address, President Tolbert, who had just assumed chairmanship of the OAU, told the delegates that South Africa and its system of apartheid was the most pressing problem on the continent.

It was a fairly stormy summit, with three walkouts in as many days. A plan for a pan-African peacekeeping force was raised and shelved. The presidents of Nigeria and Uganda shared a heated exchange, while rancorous debate over Julius Nyerere's recent invasion of Uganda and the defeat of Idi Amin's brutal dictatorship filled the hall. Several Arab countries, led by Libya and Algeria, stormed out when President Anwar Sadat of Egypt defended his peace initiatives. And Morocco threatened to resign from the OAU after passage of a resolution calling for a referendum in the hotly contested Western Sahara to determine its future.

Nonetheless, President Tolbert seemed to consider the summit a successful showcase for Liberia. After it was over, he returned his attention to problems at home. Instead of liberalizing the government, he tightened his grip. He set up a new ministry responsible for internal security and pressed through the legislature a law making striking workers eligible for prosecution. The Monrovia mayoral election, where the True Whig candidate, Francis Horton, was being challenged for the first time in nearly two decades by the increasingly popular Amos Sawyer, was rescheduled for June 1980 to allow a cooling period.

Tolbert also fired several of his cabinet members, including Finance Minister James T. Phillips. No one knew who would be appointed as his replacement, but the top candidates were me and another deputy minister by the name of Gerald Padmore. In the midst of the deliberations I had to leave Monrovia for official outside travel. Upon my return I went straight into the office to work, still wearing the jeans I had been traveling in. As I sat at my desk working, I got a call from the president's chief of staff, E. Reginald Townsend. "The president wants to see you," he said.

I told him I couldn't go right then, I wasn't properly dressed, but he insisted. "I think you'll be interested in what he has to say."

So I went to the president's office. He sat at his desk, looking weary but resolute. It had been a trying time for Tolbert, not only at home but as chairman of the OAU. All across Africa issues were popping up: only a month after the summit ended, President Francisco Macías Nguema of Equatorial Guinea, whose corrupt and brutal reign had been the subject of rancorous debate during the summit, had been ousted by his nephew in a bloodless coup. Like Amin, Nguema was a destructive despot who created a powerful cult of personality around himself. He would be tried and executed the following month, with Tolbert, as chairman of the OAU and a fellow head of state, observing the situation from afar.

I apologized to the president for my attire, but he was gracious in saying that he was not interested in what I wore but in what I could do.

"I've decided to make you the finance minister," he said to me. "Go and do your best."

That was in August 1979. Tolbert's decision had made me the first female minister of finance in the nation's history, but there was no time for accolades, no time to celebrate. Walking the streets of Monrovia, one could almost feel the tension. The Progressive People's Party kept up its meetings. The University of Liberia reopened and the students returned to class, electrified by all the riots had opened up. It was clear to everyone, surely even to President Tolbert, that things were bad.

Unless something dramatic took place, unless Tolbert managed some great swerve to avert the dangerous turns ahead, we were headed for trouble, and fast. I got down to work as quickly as I could, imposing austerity measures in an attempt to save the sinking economy and pushing an agenda of reform.

But I would have less than a year in my new job before the car went off the cliff.

CHAPTER 5

# THE 1980 COUP

I WASN'T home that night in April. I was at the house of my longtime friend; we had dined and gone to bed, and by the time the shooting started, sometime after midnight, I was deep asleep and dreaming peacefully. My friend woke me with a gentle touch on the shoulder and a note of concern shading his voice.

"Listen," he said into the darkness. "They're shooting. What could it be?"

Deep inside, though, we both knew the answer, or at least we suspected it. For months the tension had been building. In January, PAL had formally registered as a political party, the Progressive People's Party. Then, in March, President Tolbert had rearrested Gabriel Baccus Matthews, Chea Cheapoo, and other leaders of the PPP and charged them with treason after Matthews demanded Tolbert's resignation and called for a general strike to back up the demand. In the wake of the incident, Tolbert cracked down on the PPP, saying the group had been plotting the violent overthrow of the government. I cannot confirm this, but it was also reported that the administration offered rewards of up to $2,000 for the return of several party members—"dead or alive."

Now it was April—April 12, 1980—still in the early-morning hours, still before the dawn of the new day. My friend and I rose and dressed, trying to decide what to do. It was still dark outside, and the shooting would not stop. The telephone began to ring, friends calling to tell us, "They're shooting around the mansion. Something is happening."

Later we would learn that a group of seventeen soldiers in the Liberian armed forces had attacked the Executive Mansion shortly after midnight, quickly overrunning Tolbert's security forces and seizing and brutally killing President Tolbert in his bed. Tolbert's wife, Victoria, was taken prisoner, and the soldiers soon moved to free the PPP leaders, who joined them at the mansion.

One of the leaders of this action was a man of whom most of us had never before heard: Samuel Kanyon Doe. Doe was a twenty-eight-year-old Krahn from Grand Gedeh County, a career soldier who had formerly proclaimed his disinterest in politics and who had only recently been promoted to master sergeant. Now, in the space of a single, bloody night, he had become the new head of Liberia.

We turned on the radio. A man was speaking, a voice that we would come to know all too well. The People's Redemption Council, as the group called itself, had seized control of the government because "rampant corruption and continuous failure by the government to effectively handle the affairs of the Liberian people left the enlisted men no alternative," said Doe.

For the rest of the night and into the morning, the radio broadcast announcements, interspersed with African music and pop songs from America. The situation was under control, Doe said. Several junior officers, mostly captains and lieutenants but also a few noncommissioned officers, were appointed heads of government agencies. Order was restored, and the situation was under control. Flights to and from Roberts Field were suspended. A dusk-to-dawn curfew was put into place.

Hurried telephone calls reported that the scene was chaotic: soldiers driving through the streets in stolen vehicles, looting stores and shooting into the air in celebration. By dawn my name was on the air. I was being called in, as were the governor of the central bank and other members of the Tolbert cabinet. We were ordered to report to the Executive Mansion.

My friend, who was worried for my safety, asked, "What are you going to do?"

I wasn't sure; but I knew what I was not going to do. "I'm not going," I said. "Not without protection."

Once the sun rose, people had begun pouring into the street, singing and dancing in jubilation. From the window we could see the market women taunting those they felt had oppressed them for so many years: the members of the settler elite class, now largely hiding in their homes.

"Congo woman born rouge, native woman born soldier!" they chanted. Meaning that Congo, or Americo-Liberian, women might have birthed Liberia's corrupt rulers for 150 years, but now the native women had birthed soldiers, and those soldiers had taken charge. I knew that if I stepped into those streets with my complexion, no one would pause to ask who my father was, where my mother had been born, what I considered myself. I would be shot long before I reached the mansion.

I telephoned Charles Green, the governor of the central bank, who was also being called in. I asked what he planned to do.

"We have to go to the mansion," he said. "If we don't they'll come for us. They want us to open the bank and give them money."

"When you go, tell them where I am," I said. "Tell them I will come, but, given my complexion, I need protection. They must send someone for me." I don't know if Green ever delivered the message, but no one came. A few hours later my friend suggested that it was best that he escort me.

The streets were wild. As we drove through them, I held my breath. At one point the truck swerved into a gas station where a crowd was milling excitedly about. As the soldiers leapt out to fill the tank of the car, someone pounded on the roof and peered into the window, demanding, "Who you got in there?"

I knew it was not an idle question, and I knew the soldiers would not necessarily protect me if the crowd wanted blood. One of the soldiers responded, "Ellen Johnson Sirleaf! Taking her to see the man!"

As soon as they heard my name the crowd released a kind of cry of approval. Someone called out, "That's the woman who told us the truth about Rally Time!" They moved away, and I knew I was safe for the moment—ironically, because of a mistake I had made. Rally Time had been a Tolbert fund-raising campaign intended to generate $10 million for rural development. Although most people approved of the goal, many Liberians came quickly to disapprove of the methods used to raise the cash. Employees in government service and some private industries were coerced into making contributions.

Steve Tolbert also used to get fun out of sending adults and children—my son Rob included—out onto the streets to mobilize resources from drivers and call out, "Rally Time!" People wanted to know how much money was being raised and where it was being spent, and at one public event I issued a statement declaring that more money had been raised than official government reports indicated. Later I realized I had made a mistake in calculation, but by then the image of me as the Rally Time truth teller had become established. And it was that image that saved my life that day. The crowd, having established who was in the car, moved on. They didn't know my face, but they knew my name and the positions I had taken in the past.

We arrived at the mansion. Doe and his soldiers were not in the main building but outside in the palava hut. The yard was crawling with soldiers—young, wild-eyed, glittery with excitement at what they'd done and still might do. Groups of hangers-on, mostly young men and boys, milled around the soldiers, smiling and clapping them on the back and generally treating them as heroes. Some of my colleagues in the cabinet who had been arrested sat in a circle off to the side, heads down, guarded by soldiers with guns.

I believe that by the time I got there Green had already opened the central bank and given out what money was there to the soldiers. For some, perhaps even for most, it was more money than they had ever seen before, and they were jubilant. So by the time I reached the palava hut I wasn't sure what they would want of me.

Doe sat on a chair in the middle of the hut, holding something in his hands I could not see. I knew immediately that he was the one in control, though I had never before laid eyes upon him. It was the way he sat there, his chest puffed out a little, and the way the others fawned about him that told me. He looked up at me, and I saw what it was he held in his hands: a copy of the most recent budget.

"Do you know this budget?" he asked, looking at me. "Do you understand it?"

"Of course," I said. "That's my work."

He nodded, then held the budget out to me. "You'll explain the budget to so-and-so," he said, indicating the person he'd chosen to take over this particular job.

They took me aside and had me sit for two hours explaining the budget to this man, and during that time it was made clear to me that when I finished I would be sent either home or to jail—the decision had yet to be made. So I simply focused on the job at hand, telling myself to remain calm. At the end of the two hours Doe glanced my way and said, "Okay. Let her go home."

I stood up and began to leave the room. Just as I reached the door Doe called out, "Stop."

I froze where I was, unsure of what would happen next.

"Send some people with her," Doe said to a soldier near the door. "To take her home safe."

On the way out I was taken by the morgue to view President Tolbert's body. It was a terrible sight; he had been brutalized. I said a brief, silent prayer for him and his family. Then I was taken out.

Later on, when Doe became fearful and oppressive and the situation went from bad to worse, I would be accused of having supported the coup because I did not resist that morning, because I did not stand before Doe and his men and challenge the coup. But from the moment I first heard Doe's voice on the radio announcing the coup, I had decided upon my course of action. A terrible event had taken place. The collision we had so long been driving toward had finally taken place, and there was no undoing that fact. I had raised my voice as loudly and as often as I could in warning, but still we had been unable to avert the situation, and now what? The car of state was still moving, and we were all still in danger. Should I fight the person at the wheel? Or should I try to help?

The soldiers took me back to my friend's house and let me out. I thought they would leave, but they simply parked themselves in the front of the yard. All afternoon they remained there, and into the night. I had no idea why they were there or what their orders might be, but I knew I could not go to sleep without having a better grip on the situation. So I kept going outside to talk to them, to make a connection, to personalize myself. "How are you doing?" I'd ask. "Do you want something to eat? Some rice? Some bread? Can I get you something to drink?"

Finally, after midnight, I went out again and one of the soldiers turned to me and said, "Go to sleep, woman. The man didn't send us here to kill you, so leave us alone."

I did.

The next day Doe installed himself as head of state and again went on the radio to announce the formation of a new, fifteen-member cabinet made up of both soldiers and some civilians, including Gabriel Baccus Matthews as the new foreign minister and Togba Nah Tipoteh as the minister of planning and economic affairs. Doe also announced that members of Tolbert's former administration—nearly one hundred people—had been rounded up and put into military prison to await trial.

For me, the pattern of that first day repeated itself on the following ones. Soldiers would arrive in the morning, drive me to the mansion, sit me down somewhere in the yard, and ask me questions about the budget or the bank or financial information about running the government. In some ways, it was a minor form of torture; I never knew when they arrived if they had been ordered to take me to the mansion again or to the prison where my former colleagues were awaiting their fate. I never knew if, once I had answered all their questions and they had no more use for me, I would be released or simply shot. No one went out of his way to reassure me, to tell me things would be okay.

That things would not be okay, at least in the short run, was clear to all of us; we knew the country was in for a difficult time. Monrovia toppled on the edge of anarchy, with a dangerous kind of euphoria sweeping the poor and oppressed. Every day the streets filled with taunts and threats: Kill all the Congo people! They have stolen the money, they have stolen the land! Day and night drunken soldiers roamed the town in cars and trucks, yelling, laughing, making wild noises, threatening, harassing, and beating whomever they saw fit.

Score settling began. People who had worked as servants or lived as wards in the houses of the well-to-do now brought soldiers to ransack those very homes, to seize and imprison their employers or foster parents, sometimes simply to shoot them in the street.

Nor did one have to be a so-called America-Liberian to reap the whirlwind. Anyone who was seen as being educated, anyone with resources or a certain degree of material wealth, was a potential target, even if his background was indigenous. That realization frightened all of us.

In the end, what saved me during this turbulent time was my years of rebellion, my history of standing up and speaking out against the Tubman and Tolbert administrations in particular and an unjust society in general. Without that history, without that reputation, I would surely have been lost.

For my part, I kept going to the mansion as required, trying to be helpful to those I met. I tried as best I could to calm the situation, to suggest that okay, the president had been killed but now we could all move forward without further violence, to focus our attentions on getting the country to the place where they said they wanted it to be. I knew all too well that I was dealing with the perpetrators of a brutal assassination; of that there was no doubt.

But they were also part of a movement that had brought about change to our society. No matter how difficult, that change needed to come and one needed to turn it into something positive. Those of us with skills and experience could help get things back on a safe track, or we could stand aside and watch as the situation deteriorated even more.

For a while there was hope. For one thing, Doe named a large number of civilians to his new cabinet, including representatives of the PPP and MOJA. Though it was clear that real authority still rested with the original group of soldiers, now all promoted to exalted ranks and calling themselves the People's Redemption Council, nonetheless, the presence of civilians offered the hope of a moderating influence. Doe himself was promoted to general. But, to his credit, he scrupulously refrained from calling himself president, adhering to the less democratically charged title head of state all the way until the 1985 elections.

Two days after the coup, General Doe went on television to address the nation. Flanked by armed soldiers and wearing his green fatigues, an army cap, and sunglasses, with a grenade dangling from his pocket, Doe sounded almost conciliatory, saying "We are prepared to let the past go quickly into history."

He promised a new, more open and more just society, one run without discrimination based upon descent or class. "Gone forever are the days of who you know and 'Do you know who I am?'" he said. "We now enter a time of 'What can you do?'"

That same day, Doe's new minister of state for presidential affairs, George Boley, announced that the first scheduled trial of former cabinet members, in this case Joseph Chesson, the former minister of justice, was being postponed. The announcement raised hopes that international calls for the new regime to forgo the kind of brutal show trials that far too often followed a coup d'état were being not only heard but understood. The next day Doe, seeing the situation begin to spiral out of control, consulted with U.S. officials on the ground and then imposed security measures intended to stop the looting and violence. Several soldiers accused of looting and other subversive activities were executed in what Doe said was an effort to restore law and order. Though sporadic violence continued and the situation remained incredibly tense, for a while we pushed ourselves to believe. The worst was over, better would come. Liberia, once again, would be different. Liberia would show the way.

Two days later, the trials began. Former minister Chesson and several associates were the first to be taken, barefoot and degraded, before a five-man army tribunal. Like those who would follow, Chesson and his associates were allowed no defense counsel. They were given no details of the specific charges against them, being instead accused only of blanket "treason, rampant corruption," and "gross violation of human rights." Within days they were found guilty and sentenced to death.

One of the most difficult times for me, during those first, almost hallucinatory days when I had to keep reporting to the mansion, was the day I arrived and was met by Doe outside. He put on his big, dark sunglasses and strode purposefully toward his car.

"Come," he ordered. "Go with me."

He drove me through the streets, observing the city and its residents, until we finally arrived at the Barclay Training Center, the BTC. When the car stopped I saw, there in the yard, all these people I knew, government officials of this order or that. They were lined up under guard, with their shoes and shirts missing, and among them was my brother Carney.

One can imagine the embarrassment I felt at that moment, with my own brother standing on that line with the rest of the men while I stood there in the company of the person who had put them there, the self-declared head of state. I knew even then that many of those men would look at me and assume that I was a party to, and a supporter of, the terrible events taking place. I knew that, but I also knew there was no way I could have refused to get into General Doe's car. There is no way I could have refused to go with him and lived.

As we drove away, I did turn to him and say, "I don't know why you have my brother on that line. He has done nothing wrong."

Doe said, "I don't know myself."

Subsequently we found out that my brother had been put into the BTC by one of Doe's confidants and the new minister of justice, Chea Cheapoo. The reason: Mr. Cheapoo had been in the legislature before the coup but had lost his seat after only a month or so, and my brother had laughed at him for it.

Looking back, it seems clear that despite a whirlwind of calls for moderation in those first few turbulent days, Doe became afraid. He had plenty of street smarts—one must credit him with that. But he was also young, uneducated, and utterly unprepared for a job he had seized almost unthinkingly and mostly because he could. No doubt helped along by his confidants, he began to fear this particular group of well-known and well-established men. These were prominent people, whom he believed possessed huge resources. He feared if allowed to live they would surely stage a countercoup.

On the morning of April 22, ten days after the coup, Doe held a seven-minute press conference, after which reporters from all over the world were invited to the beach at the Barclay Training Center.

Nine massive wooden posts had been buried in the sand. As reporters watched and thousands of civilians and soldiers stood by and cheered, nine former members of Tolbert's administration were stripped to the waist, led to the posts, and tied up facing away from the sea.

According to the *New York Times*, so many soldiers stood around jeering and taunting the prisoners that it took an officer half an hour to move them back far enough to clear the way for the firing squad. Eventually, however, he succeeded.

The men were then shot to death in a terrible barrage of bullets. Their bodies were cut down and left at the foot of the stakes, while four other men were brought forward and tied to the posts. They too were executed.

In all, thirteen men died on the beach that day.

Those killed were: Frank Tolbert, the president's brother; Richard Henries, Speaker of the House; C. Cecil Dennis, foreign minister; E. Reginald Townsend, chairman of the True Whig Party; Joseph Chesson, justice minister; James A. A. Pierre, chief justice; James T. Phillips, former finance minister; David Franklin Neal, former minister for economic planning; Frank Steward, budget director; Cyril Bright, agriculture minister; John Sherman, trade minister; Charles T. O. King, a congressman; and Clarence Parker, True Whig Party treasurer.

I was at my mother's house when the execution took place. As I remember, the afternoon got cloudy and dark and an ominous feel-ing pervaded the town. But no one not at the beach knew what was happening, and I didn't hear about the day's events until much later that night. We had all followed the trial as best we could through the limited media being disseminated at the time. We knew the men had been tried in that kangaroo court and found guilty, but we had no idea they were going to be killed. The curfew imposed immediately after the coup was still in place, and the streets were still unpredictable. You didn't go wandering around seeking information. You stayed put as much as you could.

Only four ministers from the Tolbert government were spared. I was one of them.

Years later, as I began to challenge General Doe about some of the malpractices going on in the country and as our complex but workable relationship began to sour, Doe liked to tell the nation a story about why I had not been executed that fateful day. I had been spared, he told the world, not because the so-called revolutionaries believed me innocent of the same "crimes" committed by the murdered men but simply because of my mother.

According to Doe, he and his men had once found themselves in some area where my mother happened to be, and when they told her they were thirsty, she gave them water to drink. Doe claimed to always remember this and so wanted to be kind to her daughter when he was able to be. Had it not been for my mother's kindness, Doe used to say, I would have been number fourteen on the beach that day.

We asked our mother about the truthfulness of that statement, but she couldn't remember. We think it never happened, but the story con-tinues to circulate in Liberia. Now, when people ask if it's true, we say, "Our mother was certainly that kind of person. If someone was thirsty, she would have given them water, so let's accept the story as it is."

My brother-in-law, Estrada Bernard, Jennie's husband, was minister of labor, youth, and sports. As fate would have it, he had been in Ghana for a football game at the time of the coup and so was also spared. He did not return to Liberia, and very soon, as soon as she could, after being subjected to unimaginable acts of terror, Jennie smuggled her children out of the country and then followed herself.

The seaside executions brought a storm of international protests down upon the heads of Doe and his governing council. Whether the protests had an effect it is difficult to say; however, the executions ceased, though dozens of people connected to the Tolbert government remained in prison.

Many of those who could, such as Jennie, had fled the country, while others went into hiding. Doe occupied the Executive Mansion, declared martial law, and suspended our 133-year-old Constitution. He still had me come every morning to answer questions. Finally he asked if I would serve as president of the Liberian Bank for Development & Investment. I accepted the post.

Accepting put me right in the middle of two increasingly antagonistic groups. Some members of Doe's fledgling government distrusted me because of my former position in the Tolbert administration. On the other hand, many of those associated with the Tolbert and Tubman governments who remained in the country viewed my decision to work with Doe in starker terms: they considered it betrayal.

Nonetheless, my focus was the nation. I was still trying, naively perhaps but with every ounce of strength I could muster, to turn the situation toward the good. I thought I could help accomplish that better from the inside than the outside. What's more, the country's economy was under serious strain, and Doe and his advisers had little, if any, understanding of how to stave off financial disaster. Liberia had accumulated a significant debt due to the hosting of the OAU; the country held perhaps $5 million in the central bank, while owing upward of $700 million in foreign debt. Furthermore, in the wake of the coup the country experienced a massive outflow of capital—individuals and corporations alike began to send their money out. Foreign investors were nervous. Much of the skilled workforce soon fled. All of this exacerbated the dire financial situation of the new government. I feared the economy was on the verge of collapse and felt it was my duty to help avert such a disaster in any way possible.

But there was tremendous tension in the fledgling government, tensions between the army men in the People's Redemption Council and in the cabinet and those on the civilian side, who represented a more established element. To say that I remained at the job under serious strain would be an understatement. Even though I had long been known as a progressive agitator among the old guard—and in fact that reputation had saved my life—I was still considered by some to be part of the past. So many of the young people around Doe resented that past and anyone associated with it. Every day I was faced with increasing tension between the old order and the young progressives, who increasingly felt that all positions of authority and power should pass to them. The agitation grew greater and greater.

It was a strange time in Monrovia. In some ways, life went on as before: people woke and worked and slept, the market women still gathered to sell their wares and goods. In other ways, however, everything had changed. Doe's soldiers had taken over some buildings, including the homes of some of the thirteen men killed on the beach, and simply shot out the windows of others, such as the grand Masonic Temple built by Tubman. The soldiers drove recklessly through the streets, wrecking cars and harassing civilians as they saw fit.

Nonetheless, I pushed ahead. In June I accompanied Togba Nah Tipoteh, the new minister of planning and economic affairs, on a trip to Washington and New York, along with several other members of the new cabinet. The trip was intended to seek emergency aid, to boost the United States' confidence in the new government, and to introduce the new officials to the world financial community. Jack Clark, who was vice president of Citibank and with whom I had developed a working relationship of mutual respect, hosted a meeting at his office.

Citibank had, over the years, served as the central bank for Liberia and as the country's main institutional financial supporter. In taking Minister Tipoteh and the others to meet the Citibank officials, I hoped to use the opportunity to stabilize the teetering economy. My hope was that Citibank would use its prestige and leadership in the international financial world to assure other financial institutions that, though there had indeed been a coup, Liberia fully intended to maintain its economic policies. This would be just a slight interruption of a political nature, I proposed, but the country's economy was sufficiently strong, based upon a very good natural resource endowment, to see it through. In other words, as long as no one panicked and no one acted rashly, Liberia would be fine. To prove it, here, meet the man in charge of economic planning for this new government.

But no sooner had Tipoteh begun to speak than he began to justify the coup, in language so inflammatory one could practically feel the temperature in the room begin to climb. The problem was that Tipoteh tried to justify not only the coup, with its violent seizing of power and the loss of a president's life, but also everything done to that point in its name—including the public executions of the thirteen men. He either forgot or failed to realize that many of those same men had formed warm and cordial professional relationships with some of the people in that room. Some had even been friends.

What Tipoteh failed to realize was that in justifying their killing he only served to increase the disappointment and anger in that room and in the international community, anger at actions that need not have been taken. The truth is, the coup would have been accepted if those killings had not taken place. At the time, sitting in that room, I sensed more trouble ahead. I knew then that they were not going to really get the support they needed and that it would take a long time before the county regained the confidence of the international community. Which meant the economy was headed for serious trouble and soon.

As Tipoteh spoke, I grew more and more annoyed until finally I found myself unable to sit there any longer and continue to listen. I walked out. Mr. Clark followed me and, like Professor Papanek before him, suggested I might want to consider leaving Liberia. "Listen," he said, "if you decide you need a job, get in touch with me."

It was the first time since the coup I had seriously considered leaving. As time passed and Doe and his People's Redemption Council solidified their control of the country, it was becoming increasingly clear that the promises of the revolution would not be fulfilled.

It was a painful realization for those of us trying to walk, drag, or carry the nation to a new and better place. The truth is that Doe had a tremendous opportunity in the first days, weeks, and months after the coup. His popularity was sky high; walking through the streets of Monrovia, he was greeted by cheering, chanting crowds numbering in the thousands. Many of those who would naturally oppose his new administration had fled, and though there would be some scattered, disorganized attempts at counterrebellion in the first months, they would quickly fall apart. Doe held in his hand an astonishing opportunity for meaningful change in Liberia. But instead of grasping it, he tossed it away.

The bottom line is this: Doe got greedy. He got greedy and the people around him got greedy too, and collectively they began to feed off the state's largess like a pack of hyenas. Money poured out of the government's coffers at an astonishing rate. Any attempt to formulate a vision for moving the country forward was soon abandoned; what rose in its place was a kind of copycatting of the past. Doe, a man who before April 12, 1980, had lived in the most impoverished of tin-roofed barrack shacks (a major contributor to the coup in the first place), a once humble man who had never expressed much interest in either power or politics, soon developed a taste for chauffeured Mercedes and plush country homes. His men—young, uneducated, and before the coup mostly desperately poor—took everything they could get.

Then, when the truth of what was happening became apparent, when people began to notice what was happening and complain, the oppression began.

As it turned out, however, I was wrong about the economy—at least in part. As I had foreseen, the country's financial situation quickly deteriorated. By January 1981 Liberia balanced a monthly revenue, mostly from iron ore and rubber, of about $17 million with monthly expenditures of about $30 million, including $7 million in debt repayments. The Doe administration began lurching from month to month, trying to stave off bankruptcy: a $7 million emergency grant from the United States here, a $4 million loan from several foreign banks there.

But things never hit rock bottom, and for one simple reason: the United States held the country up. Fearful that Doe would give in to overtures from Soviet bloc nations, including Libya and Ethiopia, the United States quickly stepped forward to pull Doe into its embrace—especially after Ronald Reagan took office as president in 1981.

President Reagan considered Liberia an important Cold War ally in Africa, one that could help staunch the spread of Soviet influence and stand as a bulwark against the socialist-leaning regimes sprouting across the continent. During the 1980s, Monrovia shone brightly as a kind of American satellite. The Voice of America's main relay station, which transmits to all of sub-Saharan Africa, was located just outside Monrovia. So too was the giant Omega Communications Tower in Paynesville, used to monitor shipping in the Atlantic and said to be the tallest structure in all of Africa. Both served American interests during the Cold War, while common knowledge held that the main CIA station in Africa was based in Monrovia's massive U.S. Embassy compound.

Thus U.S. financial aid, halted in those first, confusing months following the coup, quickly resumed. Soon American money was flooding into Liberia. From a low of about $20 million in 1979, aid rose to $95 million by 1984 to a total of more than $400 million between 1981 and 1987. This was more than double the amount the country had received during the previous three decades, according to the *New York Times*.

On the first anniversary of the coup—dubbed National Redemption Day by the Doe government—the Reagan administration sent the missile-firing destroyer *Thorn* and 100 Special Forces soldiers, or Green Berets, to Liberia in a showy display of military support. The Green Berets staged a public parachute display on the anniversary itself, then ran a one-month training program for the Liberian armed forces, intended to shore up discipline among the 6,000-man force and, not incidentally, make anyone considering a countercoup have second thoughts. Doe perhaps had some affinity for the Special Forces, having himself been trained by them two years earlier. Nearly $15 million of U.S. aid during this time was dedicated to military expenditures, including new equipment and training and, importantly, construction of new barracks.

In return, Doe expelled the Soviet ambassador, chastised Libya, and established ties with Israel.

Washington also put pressure on the IMF to come to Liberia's rescue, and the IMF responded with "balance of payment" loans exceeding, in one instance, more than $85 million. Such loans are not directed toward any particular development activity but are designated as a general infusion of cash meant to stabilize a country and allow it to meet its budget. The only problem was that the Doe government lacked the capacity to handle the economic situation—it didn't know how to run an economy and the officials refused to listen to those who did. In no time at all the government simply stopped servicing the debt, which then proceeded to balloon.

When I took office as president in January 2006, this debt was still wrapped around Liberia's neck, hobbling us as we tried to rebuild our shattered country.

As of June 2007, Liberia's outstanding external debt was estimated at $4.7 billion, equivalent to 2,300 percent of exports. Much of this debt was accumulated interest from, first, the hosting of the OAU conference and, later, the bailout of Doe's government.

Of this figure, approximately $1.6 billion was in multilateral debt, owed to the World Bank, the African Development Bank, the International Monetary Fund, and other multilaterals. An additional $1.5 billion was in bilateral debt, owed to the so-called Paris Club, an informal group of official creditors from nineteen countries, and to non–Paris Club countries, including Taiwan, Saudi Arabia, China, and Kuwait. (China has announced its intention to forgive the debt outstanding.) The remaining $1.6 billion was owed to commercial creditors: commercial banks and hedge funds. Some of this was in the form of suppliers' credits, most of which probably will not have to be paid because the statute of limitations has expired.

Managing this suffocating debt has been one of the top priorities of my administration, and we have made tremendous progress in this regard. As of this writing, in early 2008, approximately $900 million of that debt has effectively been forgiven by the World Bank, the African Development Bank, and the Paris Club, so the amount outstanding is now closer to $3.8 billion.

Furthermore, Liberia is servicing its World Bank, ADB, and IMF loans through new financing from each organization, so net flows are zero. The remaining amounts will be forgiven at the HIPC Completion Point through the Multilateral Debt Relief Initiative (MDRI).

HIPC (Heavily Indebted Poor Countries) is an initiative, launched in 1996 by the World Bank and the IMF, to provide low-income countries struggling to cope with heavy debt a chance to make a fresh start.

In return for debt relief, participating governments pledge to implement key economic, legal, financial management, and development reforms and provide specific, detailed plans to reduce poverty and improve the quality of life for poor citizens.

The decision point for participation is reached when a country makes this pledge to reform, establishes a track record of macroeconomic stability, clears any outstanding arrears, and prepares an exhaustive paper for the World Bank describing the key structural and social reform proposed. Liberia reached Decision Point in March 2008.

To reach the Completion Point, a country must satisfactorily carry out those key reforms for one year and maintain macroeconomic stability. The amount of debt relief then becomes permanent. Liberia could reach Completion Point in the second half of 2009.

In regard to our bilateral debt, in April 2008, Liberia reached an agreement with the Paris Club leading to an immediate reduction of $254 million. The remainder was rescheduled over many years, with zero due for the next three years. This was the most generous treatment ever provided by the Paris Club at HIPC Decision Point. Most of the Paris Club members also announced their intention to forgive 100 percent of amounts owed at the HIPC Completion Point.

Our government has also been meeting with private creditors, and there are positive signs that an agreement can be reached with them on a debt buyback.

* * *

DEBT IS, of course, a tremendous problem throughout Africa, and in many cases the source can be traced directly back to the first years after independence. Much of the debt incurred by many nations came not only via military buildups but also through bilateral loans. The problem was that not enough care was taken to make sure these loans were directed into productive activities—creation of industry, expansion of agriculture—that would generate repayment of the debt. Instead, the loans were tied to politics. This was at the height of the Cold War, when the competition between the United States and the Soviet Union for spheres of influence was fierce. All a head of state had to do was declare his loyalty, and the money flowed in.

If that head of state and his government had the capacity to properly manage the debt, to funnel the money into activities that would generate income, things worked out. But if, as in the case of Zaire, the loans were used solely to prop up a failing government, the foundations were laid for a serious debt problem. That is what happened in Liberia.

Seeing this, I knew I once again had to speak out. At a graduation ceremony for the Booker Washington Institute in November 1980, I stood before the audience and criticized what I saw happening, which was essentially a re-creation of the same old story, only this time with new characters in the starring roles. I told the story of the rat trap in the house, in which a farmer sets a trap for a rat that has been stealing his rice. But the rat is too smart to be trapped; instead one day the farmer's wife accidentally steps into the trap and injures her foot. Gangrene sets in, and the farmer has to kill his chicken to make soup to try to save her. She dies nonetheless, and the farmer has to kill his cow for the feast for her funeral. I told the story as a metaphor for the coup: in trying to kill the rat the coup leaders might also end up killing other things, things far more precious to the country and even vital to the country's survival.

Thomas Weh-Syen, vice chairman of the PRC and Doe's second in command, was told about the speech. Word came back to me pretty quickly that he had not appreciated my warning, and I started getting messages that he was setting for me. I knew then I had to leave.

I was in the unique position of still being, officially, a governor and employee of the World Bank only seconded to the Liberian government. It had never happened in the bank's history, and afterward it changed the rules so it would never happen again, but it was very helpful to me at the time. I contacted my superiors at the World Bank in Washington and asked them to recall me to my post. The people at the World Bank, of course, knew what was happening in Liberia. They quickly wrote to Doe and asked him to release me. He was still trying to be a good boy on the world stage at the time, and they put him into a position where he could not say no. I was released. Later, of course, Doe killed many of the very PRC members who had been his closest allies, including Weh-Syen.

Before anyone really knew what was happening, I packed up my things and returned to Washington. My sister, Jennie, who had left Liberia right after the coup, was already in the United States, as were three of my four sons, struggling on their own to go to college and build their lives. They all breathed a sigh of relief to see me. It was December 1980, eight months since the coup. For the first time I left home with a heavy heart, not knowing when I'd be able to return or what I might find when I did.

# CHAPTER 6

# CLIMBING THE CORPORATE LADDER

---

I REMAINED at the World Bank for less than a year, but it was during that time that I had the opportunity to meet Robert McNamara. It was a brief meeting, right after I returned to the bank, but the forcefulness of this man was not lost on me. We made a connection and have kept in touch over these many years.

McNamara, the U.S. secretary of defense under President Lyndon Johnson during the Vietnam War, had come to the World Bank in 1968. Under his leadership the bank tripled its staff and increased its lending nearly sixfold, with the greatest increases accruing to Africa and Latin America.

To some he is a hero for expanding the bank's mission to include the eradication of poverty through a process of integrated rural development and instituting an environmental dimension to lending practices. It was during his term that the World Bank launched its largely successful campaign to control onchocerciasis, or river blindness, a terrible parasitic disease spread by the bites of small black flies. However, some critics say the bank's most remarkable feature during his tenure was its tendency to reward dictatorial regimes with generous funding as long as they were anti-Soviet and anti-Communist.

For my part, I believe Mr. McNamara's commitment to Africa was well-intentioned and utterly sincere. I said as much in public once. After I'd left the bank I was appointed a member of its first Council of African Advisors. It was during a meeting of the council that Mc-Namara made remarks in which he stressed how much the bank had increased its funding and attention to Africa during his tenure. Ali Mazuri, a prominent academic and political writer and member of the council, grew quite upset during the speech and afterward took Mc-Namara to task for, in Mazuri's words, saying he had "saved Africa." I went to bat for McNamara. "It's not a question of him saying he has saved Africa," I said, "but of him saying he's totally committed to Africa." I believe he was.

But as much as I enjoyed and appreciated the World Bank, there came a time for me to leave it. About a year after my return Citibank, via Jack Clark, offered me the chance to become the first African woman appointed as a vice president. I accepted and was assigned to the Citibank African regional office in Kenya.

In Nairobi, I was in charge of Citibank's "nonpresence" countries, i.e., nations where the bank had yet to establish itself. Thus I spent a lot of time traveling to countries in eastern, southern, and central Africa, such as Uganda, Rwanda, and Ethiopia, where it was my responsibility to market Citibank products. This proved very valuable to me, as it enabled me to meet and befriend the leaders of these nations—finance ministers, central bank presidents, and presidents themselves. These contacts would be very helpful later on in my political career.

In particular I recall the relationship I developed with the leaders of Uganda. During my time in Nairobi, Uganda was going through an intense conflict, with President Apollo Milton Opeto Obote being threatened by a bush-based guerrilla rebellion led by Yoweri Museveni. I visited Kampala several times, sometimes with World Bank Vice President Willi Wapenhans, and each time found it in a devastated state because of the constant attacks. Although I was meeting primarily with financial officials, I did have the occasion to meet President Obote and found him to be a man whose very demeanor intimidated those around.

Even as I became good friends with many of his cabinet members, including the then minister of finance, it was clear to me that the basis was being laid for the fall of that government. Rumors were swirling that government oppression was getting worse, that people were being killed in the night in the basement of the presidential palace. At one point I told my friend the finance minster, Ephrahim Kamuntu, "The writing is on the wall. It will not be too long before this government will not last." Our own experience in Liberia showed what happens when a government ignores an increasingly radicalized society or responds with more oppression, I told him, but he dismissed my warning. Many years later he would say I had been right: they had missed the warning signs. Clearly they should have listened more to the internal warnings of unrest.

After Museveni succeeded in pushing Obote from power, Obote fled into exile in Tanzania and later Zambia. Museveni became president, and I was introduced to him. I found him to be a man who was very forthright and spoke his mind clearly and without hesitation. He possessed all the revolutionary characteristics we all had or admired and wanted to have. We became good friends.

In some ways, working for Citibank was a whole new ball game. For one thing, the compensation was wonderful. I had a very comfortable life: a big home in Muthaiga, lovely neighborhood north of the city center that is known as Nairobi's Beverly Hills, plus a chauffeured car, domestic servants. My mother came from Liberia to spend several months with me, and she enjoyed it very much. We both found Nairobi to be a vibrant, energetic city, always bustling with international conferences and the like, which drew an interesting mix of people from near and far.

Another good thing about the institution was its emphasis on learning. As a Citibank employee, you never stopped learning—there seemed to be a never-ending series of workshops and classes and courses designed to enhance our marketing skills and analytical abilities. This was wonderful for me, as I was always eager to sharpen my professional skills further.

But working for one of the world's largest financial institutions was also very demanding. To succeed, one had to become part of the

Citibank culture, the heart of which was a constant and tough-minded aggressiveness toward work. Everything centered on what you contributed to the bottom line, and it was important to remember that. One was required to write monthly reports that basically justified the good compensation and the good life you were being allowed to live. Still, I made good friends at Citibank, some of who remain close to this very day. If I need advice or support, I can call on them at any time.

Even as I was climbing the corporate ladder, though, my mind was always at home. Although all my sons were now living abroad, as were Jennie and her family, I remained connected to Liberia.

I kept abreast of developments in Liberia. Starting in 1981, Samuel Doe reported the first of what he said were planned or attempted coups. That August the government executed five members of the ruling People's Redemption Council for allegedly plotting to kill Doe, including Thomas Weh-Syen, the man who had driven me from Liberia with rumors that he was out to get me. The following year, Doe arrested, then released, after protests, several student leaders accused of disobeying the ban on political activities.

Meanwhile, Doe gave up his camouflage uniforms and dark sunglasses in favor of three-piece suits. After receiving an honorary doctorate from the University of Seoul in South Korea, he began referring to himself as Dr. Doe. Financially, massive amounts of aid from the United States kept the country afloat, but still the economy staggered.

Nonetheless, U.S. support for Doe remained high. In August 1982, Doe visited President Reagan at the White House, where he was warmly greeted by the president as "Chairman Moe."

Back home, however, other civilian leaders who had remained in the country to help the young government were also walking—or falling— away from the Doe government. Tipoteh resigned his post as minister of planning and economic affairs while on a foreign mission in August 1981. Matthews, after a falling out with Doe, was removed as foreign minister. Meanwhile Doe, like Tolbert and Tubman before him, began packing the government rolls with friends and family, in this case most of the new employees being from the Krahn ethnic group.

During this time I stayed in touch as often as I could. During my time in Nairobi, when I visited home, I would sometimes, at General Doe's request, visit him at the Executive Mansion. These were courtesy calls, a kind of checking in, which he appreciated. Doe had a lot of affection for me and even trusted me, in some strange ways. After Tipoteh left as minister of planning and economic affairs, Doe called me in the United States and offered me the job. I declined but recommended someone I thought equally capable, Dr. Byron Tarr, then alerted Byron to the appointment. That Doe later appointed him as planning minister spoke to the level of respect in which he held me. Later on, when things got bad again, I would wonder about this respect and affection—wonder if they were enough to save my life.

During those early days, whenever I made a courtesy call on Doe, he would often ask me to bring him something the next time I traveled between Liberia and America. Once he said, "When you go on your trip next time to America I want you to bring me something. I want those trousers that have pockets here and here," he said, patting his thigh front and back. "I also want that medicine you can pour in a glass that says 'Voo!'"

So the next time I came home I brought him his khaki pants and several boxes of Alka-Seltzer. For me that was a small thing to do; I thought it was better to keep the lines of communication open than to close them, to continue trying to nudge Doe toward some higher ground.

Unfortunately, some people will not be nudged.

Once, as the situation was deteriorating, I remember saying to him, "General Doe, you promised the people so many things, and you are not keeping those promises."

Doe just looked straight at me and said, "I didn't promise them shit."

## CHAPTER 7

# THE 1985 ELECTIONS

In the summer of 1983, General Doe announced a timetable for returning Liberia to civilian rule. He was under tremendous pressure to do so, not only from the country's radicalized young society but from the international community, including the United States. Doe announced that the ban on political activities would be lifted in April 1984, followed by elections in January 1985, with the new government taking office that following April, just in time to celebrate the fifth anniversary of the coup.

The remainder of 1983 passed with increasing tension. General Doe, surrounded by mounting rumors of potential coups, continued to consolidate his personal power at the expense of the Liberian state. He had a falling-out with Thomas Quiwonkpa, one of the top leaders of the 1980 coup, and transferred Quiwonkpa from his position as commander general of the armed forces to secretary-general of the increasingly sidelined People's Redemption Council. By the end of the year, Quiwonkpa fled the country, along with many of his supporters. Many said he had gone to the United States, along with two men whose names would later resonate across Liberia: Prince Johnson and Charles Taylor.

The following summer, Doe did indeed lift the ban, allowing political parties to register with the newly formed board of elections. Nearly a dozen parties sprang up to be registered. Baccus Matthews formed the United People's Party, while Dr. Amos Sawyer formed the Liberia People's Party.

Doe also dissolved the People's Redemption Council, which by this time had become largely subordinated to Doe's personal rule, and replaced it with an interim National Assembly, designed, he said, to ease the transition to civilian rule. That same summer a new constitution, drafted by a National Constitution Commission chaired by Sawyer and including Winston Tubman, among others, was overwhelmingly approved by voters in a national referendum.

All these were hopeful signs. Except that Doe had no intention of actually allowing democracy to bloom in Liberia.

When the new constitution angered him by setting thirty-five as the minimum age required to be president, he simply added two years to his own age to make the cut. When so many political parties sprang up to be registered, Doe became more oppressive instead of less, passing the infamous Decree 88A, which made it a crime, punishable by ten years' imprisonment, to spread "rumors, lies and disinformation" about the government. He established a Special Elections Commission to run the election and stacked it with loyal partisans. He announced his intention to stand for the presidency despite years of promising to "return to the barracks" and formed the National Democratic Party of Liberia. The NDPL easily managed the registration process, but other parties had more difficulty.

Nonetheless, hope once again sprang alive. I began to receive telephone calls from friends and others who felt it was time for us to form a political party to challenge Doe's military dictatorship and urged me to return home to help. I decided I would. That July, during my vacation from my job at Citibank, I flew back to Monrovia to become one of the founding members of what was then called the Liberian Action Party. The plan was that Jackson F. Doe, a popular politician who was no relation to Samuel Doe, would stand for president and I would stand for vice president.

It was not a difficult decision, though perhaps it should have been. As it turned out, there would be consequences, to which I did not pay much attention at the time. I was working for a private financial institution in which one was not supposed to get publicly involved in politics. But I knew both that my prestige, credibility, and popularity in Liberia would add to the strength of this new political party being formed and, crucially, that only a strong party would be able to mount a credible challenge to Doe. So I agreed.

However, it took little time after I arrived for me to understand that Doe had no intention of allowing our party, or many others, to be registered. I left the country. I had been invited to Philadelphia, to a meeting of a group called the Union of Liberian Associations in the Americas. The meeting was to mark Liberia's Independence Day, and I was asked to give the keynote speech.

As usual, I spoke my mind.

I opened by telling the audience that I had just spent the past three weeks at home and that, in that time, I had felt a strong sense of destiny in the making—for better or for worse. On the one hand, there was a growing tenseness so reminiscent of the tenseness that had gripped the country in 1979 that even the brave had to be afraid. On the other hand, there was a beautiful kind of excitement. The political consciousness gripping the land was palpable and unlike anything the country had ever seen. Never before had the workings of government been so discussed and laid bare. Never before had the shortcomings of the government been so ridiculed. Never before had the desire of the people to participate in the running of government been so pronounced.

However, for all that excitement, four aspects of the Liberian agenda needed to be addressed. First and foremost was economic recovery. I explained to the group that Liberia had survived decades of wanton milking of the public purse—"eating money," as it is called—only because the prices for our two primary imports, rubber and iron ore, had remained high. Those days were over now, yet the milking continued at an unprecedented level, even though the economic pie was considerably smaller and the national capacity considerably weakened.

If the nation were to survive and prosper, the government would have to reduce the level of its intervention in the economy and, as a result, the amount of bribery and kickbacks flooding the land. What's more, it had to stop pouring national resources into nonproductive investments. The major economic activity taking place in Liberia at the time was the construction of public buildings—the Ministry of Health, the looming Ministry of Defense, and scores of markets everywhere. Meanwhile, agricultural and rural development programs slowly withered on the vine for lack of funds. Nor was the government encouraging the creation of an indigenous entrepreneurial class. Then—as now, unfortunately—most small and medium-sized businesses in Liberia were run by foreigners, particularly Lebanese businessmen. I did not then—and do not now—mean that Liberia should in any way discriminate against these businesspeople, many of whom have lived in the country for years. But the truth was that much of the profit generated by foreign-owned businesses was not plowed back into the Liberian economy but flowed out of the country and back to the businesspeople's country of origin. If sustainable growth were to be achieved in Liberia, it was primarily Liberians and Liberian businesses that must achieve it.

None of these reforms would succeed, however, I told the crowd, unless the management class in Liberia was improved.

"I look at this cross section of Liberians in this room today. I look beyond this room to several cities in several countries. I look at the many walking the streets of Monrovia or sitting quietly in an unbusy office (in low profile, as they call it), and then I look at the many idiots in whose hands our nation's fate and progress have been placed, and I simply shake at the unnecessary and tremendous cost which we pay under the disguise of righting the wrongs of the past."

There was a great deal more to the speech—some four pages, in fact. I stressed the need for the nation to continue bridging the historical social gap that had divided us for so long. I challenged the myth that democracy was a luxury in Africa, inappropriate to our traditions and foreign to our heritage. I said that Africans could see through the attempts by our leaders to halt the development of political institutions merely to perpetuate themselves in power, and I quoted former U.S.

Ambassador to the United Nations Jeane Kirkpatrick: "A government is not legitimate merely because it exists."

"It must belong to the people," I said, "respond to the people, acquiesce to the wishes of the people. Any government which refuses to be and to do these things must be removed by the people." I spoke very specifically about the political situation at home, telling the crowd that nearly half of the political parties seeking registration were encountering barriers, and thus thousands were being systematically disenfranchised. Finally, I called again for the reexamination of the little things that held big symbolic significance: our flag, our motto, the names of our cities and streets and ourselves.

All of this I said, but in the end, it was really one word—"idiots"—that got me into trouble. That was the word that did it. That was the word they really, really did not like.

After the speech I could simply have returned to Nairobi. Instead—and I'm not one hundred percent sure why—I decided I would fly home to Monrovia first, to visit my mother again and to check the political situation as the country moved, however haltingly, toward elections at the end of the year.

When I reached Monrovia, however, a surprise was waiting for me. The Liberian ambassador to the United States at the time, General George Toe Washington, was in Philadelphia and had reported back to Monrovia on the speech. A few days after my arrival home I was summoned to appear before General Doe. The soldiers who came also demanded my passport.

"President Doe sent for you," they said. "You must come to the mansion."

Of course I had no choice but to go. At the mansion Doe kept me waiting for more than an hour, the better, I suppose, to raise my anxieties. Finally I was called into his office, where I was met not only by Doe but by several of his generals and top ministers.

Doe sat fidgeting at his desk. I could tell he was unhappy.

"I heard you gave a speech," he said.

I did not deny it; there was no point in doing so.

Doe turned to his minister of foreign affairs, T. Earnest Eastman, a long-tenured foreign service official, who stood nearby holding some papers in his hands. "Read the speech," Doe ordered. Eastman complied.

Looking back now, the situation is almost comical, but it was far from funny at the time. Eastman began reading the speech from the beginning, adding his own emphasis on certain spots, his voice steadily rising. When Eastman reached the line in which I had said, "All the nicest words and the biggest promises written by someone trained in the use of the pen and read by someone trained in the use of the gun are taken for what they are—words," Doe cut him off.

"Oh! So you are saying we don't know book!" Doe yelled. He was on his feet now, furious and no longer hiding it. "We just know guns! That's what you are saying?"

He turned to his cronies. "They think I'm stupid! They think I don't know book!"

By this time in his reign Doe had become acutely sensitive about his lack of education. Without meaning to, I had struck a raw nerve.

Then Eastman read the line about the country being run by idiots—meaning the very men who now surrounded me in that room.

"Oh, you think we are idiots!" yelled Doe.

That I was in serious trouble this time became crystal clear.

In their fury the men in the room began to taunt, curse, and berate me. Some of their taunts cannot be repeated here, but among the tamer ones were these: Red-faced, stupid woman! Congo woman! And so on. The gist of the yelling was this: Who, exactly, did I think I was to challenge them? They were in charge, and if I didn't like it they would get me. Others had learned, and so would I.

Doe, now prowling the room, accused me of plotting against him, of working with some group of unknown others on a plan to kill them all. I denied it, of course, but he was in no mood to listen to anything I had to say, and most of his men were urging him to take care of me once and for all.

Through all of this I sat placidly, letting them spew their venom into the air. I remained calm, I never panicked. Looking back, I think perhaps that sense of steadiness I managed to project also served as a kind of deterrent to those raw and angry men. Through all my dealings with General Doe, I think he always wondered why I never cowered before him. It was almost as if I could see the question spinning inside his mind: What does she have behind her that makes her so brave?

Whether they thought I had superpowers like the witchcraft they believed in or whether it was simply the superpower of the United States they feared, I don't know. Whatever it was, it kept Doe from doing to me what he had done to so many others before.

There was only one person in that group, a Krahn general named Henry Dubar, who tried to go against the tide of anger and vitriol and to caution Doe. In the midst of all the yelling and taunting and pounding of fists, Dubar, speaking calmly, turned to his boss and said, "Chief, be careful.

"Be careful of these kind of people," he said. "You pay attention to her and you do something to her, you will just make her a hero. It's best to leave her alone. Let her just leave the country and go back to her job."

But Doe was in no mood to listen to that kind of reasoning. Instead, he turned to his minister of justice, Jenkins Scott. (Scott became one of the most dreaded men in Liberia and is now destitute and depressed. I reach out to him with some help from time to time.)

"Take her and put her under house arrest," he ordered.

Mr. Scott took me to my house in Congotown near that of my mother, taunting me by saying that I was stupid to have fallen into the trap set for Matthews or Tipoteh.

The soldiers took me home and remained there, keeping watch. It was August 9, the start of yet another political crisis for Liberia.

Ten days later Doe broke off an official trip to West Germany and swept home to announce that yet another coup had been intercepted; this one, he said, spearheaded by Amos Sawyer. Doe arrested Sawyer, along with several other prominent politicians, and put him under house arrest. In a speech to the nation, Doe accused Sawyer of trying to force him into resignation through the creation of "chaos and confusion among the people.

"After forcing my resignation, overthrowing the government, and making mass arrests of government officials and other citizens, Dr. Sawyer and his supporters had planned to install a socialist republic in Liberia with the aid of foreign countries," Doe said. "I feel personally disappointed and surprised at the involvement of Dr. Amos Sawyer, a personal friend of mine who seems such an innocent professor, in being implicated in the plot against the state. His prominent role and sincerity of purpose in preparing the constitution now stands in question at these revelations."

Sawyer's arrest sparked massive student protests at the University of Liberia. When the students refused to stop their demonstrations, Doe ordered the university closed and the administrators fired. Then, in an eerily familiar scene, he ordered his minister of defense, Major General Gray D. Allison, to go "and make them move or be moved." By most accounts the armed soldiers stormed their way through the university, stripping, flogging, raping, and eventually firing upon students. Although the Doe government denied any deaths, student leaders at home and abroad charged that more than fifty people had been killed.

Meanwhile, I remained under house arrest, not allowed to leave my home or receive visitors. Even my mother, who lived right next door, was not allowed to walk through my door. She was allowed to send me food and other supplies, however, and I could sometimes catch a glimpse of her as she walked past the house.

Sometimes I think about how frightening the events of my life must have been to my mother. But she never broke down, never crumbled, never told me to stop doing what I had to do. She was a very strong woman, and she believed with all her heart in the power of prayer. She prayed for me unceasingly. I believe I got my strength from her.

As time passed, I began to sneak out messages to my supporters throughout Monrovia and the world. One day one of these messages was intercepted. I had written, "To the people: Stand firm! Don't give in! Our course is right and we will win." Such sentiments did not go over very well with Doe. He ordered me moved from house arrest and taken to the infamous Post Stockade at the Barclay Training Center— the military prison—where nine students from the university were already being held. It was the first time in my life I was locked away in a jail cell, imprisoned against my will. It would not be the last—or, by any means, the worst.

I was placed in confinement with another young woman, a university student named Lucia Massiley. The two of us were placed in a small room with a makeshift toilet and little else. Across the large, open hallway was the main section, where the male university students were being held in one large room that had been sectioned off into smaller cells. We could not see the men, but we could hear them. They spent the day singing and doing mock radio shows and otherwise making noise so we could hear them and keep our spirits up. We weren't given anything to read or otherwise occupy our time; we spent the day talking, napping, listening to the men, and waiting to eat. Food came in once a day—rice, rice, and more rice, sometimes with a little palm oil if we were fortunate.

There were, however, some good soldiers, who would sneak packages of biscuits or other goods brought by our families in for us. They would hide them in the tops of their boots, then, when no one was looking, slip them under the door to us.

But for every kind soldier there was another who enjoyed seeing people in pain, and there was always the threat that some Doe commander or officer would arrive and decide to have some fun torturing you. We were all particularly afraid of Colonel Pennue, a PRC member and one of the original participants of the 1980 coup. Pennue had carved on his car license plate "IKT"—meaning "I killed Tolbert."

Some prisoners who arrived were made to do what the guards called "swimming" in order to reach their cells. This involved removing your shirt, then crawling, chest down, across the long expanse of sand and rocks that led to the prison area. It was really bad. People would get up at the end with their chests bleeding and raw. (Today in Liberia we are turning the Post Stockade into a museum that will serve as a reminder of that bitter piece of our country's history and the political persecution it caused. We must remember the famous warning about those who forget the past.)

When word got out that Doe had moved me from house arrest to an actual prison, the pressure on his government increased. Nonetheless, he followed through with his threat to try me for sedition, placing me on trial before a military court.

But General Dubar had been right: by arresting me instead of simply letting me leave the country, Doe had turned mine into a cause célèbre. Suddenly I was a hero; my detention and trial brought international attention and the kind of pressure Doe surely did not want. Citibank went to bat for me, demanding my release, as did my network of friends and colleagues in Washington and at the World Bank. When Doe went ahead and charged me with sedition and put me on trial before a military tribunal, the pressure only increased. The U.S. Congress got involved, passing a nonbinding resolution calling on the American government to block all foreign assistance to Liberia if Doe failed to release all political prisoners. The *New York Times* and other international newspapers began to cover the situation closely.

It is probably fair to say that the Reagan administration was not pleased that I had stood in America and insulted Doe so publicly. Having dumped so much money and so much support into his reign, it wanted him to succeed, and his failure to justify that support had not, at this point, caused the administration to give up on him. Nonetheless, once Congress got involved and began to apply pressure, the administration also interceded on my behalf, temporarily suspending some $25 million in aid to Liberia until Doe released me and the other political prisoners.

Perhaps most critically—and without a doubt most movingly to me—the women of Liberia came to my defense. For the first time the country could remember, women's groups, led by my friend Clave, collected more than 10,000 signatures calling for my release and saying they were prepared to demonstrate. This was the start of a very special relationship between me and my fellow countrywomen, the women of

Liberia. Again and again over the years they would rise to support me, and I will always stand to support them.

The military tribunal called to sit in judgment of me met at the Temple of Justice on Capitol Hill. Every day the soldiers would come to my cell at Monrovia's Central Prison to escort me to court on foot. It was during that time that I really became a kind of folk hero. The crowds would line the streets as I walked to my trial, surrounded by policemen. The atmosphere was far from menacing; some days it was almost festive, in fact. Even the policemen enjoyed it. They would walk alongside me, smiling and waving to the crowds as if they were heroes themselves.

The trial lasted about a week and consisted primarily of government representatives reading the long list of charges against me. What it came down to, essentially, was that I had used the word "idiot" when referring to those in our government. That was it, really. That was, in essence, the whole basis for the sedition charge. Given which, there was little doubt of the eventual outcome.

I was, at one point, allowed to make a statement, and I did, stating as clearly as I could that I did not feel I had committed any crime or made any attempt at insurrection, which is what sedition is all about. I said they should consider what they were doing, as Liberia needed all of its people and all of its talents to pull it into a better day. It was not, I said, in the best interest of the country to harm me or the other opposition leaders or to send us to prison for many years. I asked for more overall tolerance. I wasn't defiant in my statement, but I wasn't apologetic either. I appealed to General Doe because at that point the tribunal had declared me guilty and sentenced me to ten years of hard labor. All that was left to decide was where I would serve my time.

I had known from the moment of my arrest that I would be found guilty and sentenced to prison, but the ten-year part was a surprise. Hearing the sonorous, self-important voices of the judges as they announced my punishment I thought, Lord, how do I get out of this one? What's going to come to my rescue this time?

I had no idea. But we in my family were all so deeply religious people because of my mother that I knew where to turn. Through it all, whether in the small, close cell at the Sheflin camp or the dank, cold room at Central Prison, whenever I was in trouble, whenever I was in a dark moment I would close my eyes and the words my mother used to quote from the Bible would come to me: "Be still and know that I am God."

Those words gave me strength, always. Each time they came to me, each time I thought of my mother, somewhere deep inside a feeling would grow that made me know I would survive. Somebody up there was taking care of business, and I knew that to be true.

Eventually the tribunal decided I would be sent to the Belle Yalla compound, a notorious rural prison in Liberia, a prison from which people did not return. I knew if I went to Belle Yalla I would die.

However, this proved to be another misstep on Doe's part; the harshness of the sentence and the infamy of Belle Yalla only served to intensify pressure on the government, both nationally and internationally. At home and abroad, institutions both private and governmental called for Doe to release me and all other political prisoners being held.

Strangely, even in the midst of the trial I was never really afraid of what might happen. There would come a time, and very soon, when I would fear deeply for my life. But this was not that time.

Sitting on the concrete floor of my cell at night or watching the show tribunal during the day, I sometimes felt anxious or frustrated or even angry, but never that this was it, that I was trapped, that Doe would win this time. Deep in my bones I knew the situation would resolve, that things would be all right in the end. My faith bolstered me, as did the outpouring of support; no prison walls could hold that back. Actually, in some ways I was more concerned about what Citibank's reaction might be to having my name splashed all over the international news. By getting involved in politics I had, inadvertently but assuredly, violated the bank's code of conduct. Although Citibank went to bat for me and valiantly, it also let it be known that once I was free I would have to leave my position. My career at that great bank had come to an end. Jack Clark would never forgive me for ending a career in which he had invested so much.

I never went to Belle Yalla. Two weeks after my guilty verdict I was sitting in my cell one morning when word arrived like a bolt from the blue: General Doe had granted me and the students clemency. We were all free to go.

My release from prison shows the power of the public, the power of the people both home and abroad. Public opinion matters; if it is pointed, focused, and intense, it can turn things around. In this global age individuals are sometimes tempted to believe they have no power, not even collectively. This is not true. The public can make a difference if it is willing to take a position and stand up for a cause in which it believes. Against a united and committed public, even the harshest of governments cannot stand, for certainly this was a harsh military government against which no one thought they could do anything.

Sometimes people ask me if I regretted calling Doe and his cronies idiots. After all, I paid a high price for essentially one little word. The answer is no, I don't regret it—not then, not now.

For one thing, it was true. As I've said, Doe started off with some promise, but he quickly came to act much like the worst kind of bullying, greedy village chief. What the big chief wants, he gets. There was no accountability. If he wanted money, he simply called the Ministry of Finance and got it. It was so bad it might have been hilarious, had we not been talking about what was supposed to be a modern state. Had we not been talking about people's lives.

Moreover, the inability to tolerate criticism is a troublesome trait in any human being, but in a leader it is especially so. If you are a leader, you are going to get hit. You are going to get hit verbally sometimes with some very harsh words; you must be prepared to take criticism, to stand still and let it just pass over you without resorting to retribution or revenge. Being able to "take it" is part of the price of leadership, particularly in a country that is still not fully institutionalized, a country that, having only recently emerged from militarism and conflict, is still proceeding toward a democratic path.

Leadership has many challenges, and those challenges are serious. Too often people, in their eagerness to stand and shout "Follow me!," neglect to consider the downside. They do not, for example, consider the possibility of ostracism. But the truth is that if you want to lead and be hailed, you must also be prepared to be ostracized, because surely it will happen at some point in your career. I've gone through too many periods of ostracism to keep track of them all. Whenever I got into trouble with one government or another, I knew that many people in my life, even some friends, would abandon me. They were afraid—afraid to be associated with me, afraid they would get into trouble themselves. This is the way of the world, of human nature, and if you want to lead, you have to accept that there will be conscious attempts to push you into oblivion. You have to be prepared to be very lonely sometimes.

In a place like Liberia, with its African tradition of community and collective good, such shoving from the herd brings with it not only an emotional but a financial sacrifice. This was the painful lesson my mother and father learned all too well. Once you were an integral part of a close society, part of the sharing and caring; now, suddenly, you're all alone. The only thing you have left to rely upon is your immediate family. If you come from a good family, as I do, a very supportive family, it eases the pain. If not, you can succumb to the loneliness. You can just give up.

During these events I also felt myself blessed at being a professional and not just a politician. Each time I could fall back on my professional skills; I could go into the international community, where I had already established a certain amount of professionalism. Each time I fell out of favor with the governing powers, there was a way to crawl backward a little, to get a job, take a breather, recharge, rest, and wait for the opportunity to reengage. Otherwise, I might have been finished the first time I upset the Tubman government by speaking the truth.

At any rate, I came out of prison to a changed political landscape. I had come home in the first place to be the vice presidential candidate for the Liberian Action Party. But while I was in prison, Doe sent for members of my party and offered them a stark choice: remove me from the ticket, or the party would not be allowed to register.

"You must expel her from the party," Doe is said to have demanded. "She must be thrown out."

The party leadership acquiesced, even going so far as to release a statement condemning my "seditious" activities. Just like that, the LAP was allowed to register. Just like that, I was out in the cold.

To say that I was disappointed would be fair, and I took to task those who had abandoned me, but I did not get angry. What would have been the point? The deed was done. To some extent I even understood the pressure the party had been under. Doe had blatantly banned two popular parties—the United People's Party and the Liberia People's Party—from even appearing on the ballot. There was no reason to think he would not follow through on his threat to do the same for the LAP.

What's more, many of my friends and colleagues, not to mention my family, were genuinely concerned for my safety. One of the messages that had been passed to me in prison read, "We'd rather have a live ant than a dead elephant."

Doe and his collaborators even went so far as to attempt to negotiate with my family. On the day of my release from prison, government negotiators tried desperately to obtain my family's promise that I would withdraw from politics and quietly slink from the country. My brother said, "We cannot speak for her. We cannot make promises on her behalf because she's too strong a character."

So I understood the pressure being applied even to those on the outside. Still, I felt grateful to a few of the party members who had held firm in the face of General Doe's intimidation and lobbied against expelling me. When the party's executive committee decided to take that step nonetheless, a small group of supporters from Montserrado County, led by John Flamma, decided they would run me for senator instead. I accepted the nomination. On the day of my release, enraged at the raw attempted arm-twisting of my friends and family, I immediately announced the start of my campaign.

A few days after my release General Doe sent for me. When I arrived at the Executive Mansion I found him in his office, looking more tired than I had seen in the past. It was clear that the pressure was mounting.

"I want you to leave the country," he said. "Just leave quietly. I will return your passport, and no more charges will be filed if you will leave before the elections take place."

I told him I could not do what he wanted. Now that I had already suffered the indignity and danger of prison for daring to even consider a political campaign, I fully intended to follow through with one. He was not pleased, but he let me leave.

It was the end of September, and the elections were scheduled for October 15. We had only two weeks to campaign, but we did so vigorously, traveling throughout several counties, talking to people in their homes and in the market, on the streets and at meeting halls. I campaigned not only for myself but for Jackson F. Doe, a good man who would have been a good president. When they dumped me for vice president while I was in prison, they chose Emmanuel Kroma. I campaigned for him too, dismissing the explanation that once we won he would be made chief justice and I would serve as vice president.

By this time the international spotlight was shining so brightly on the country that Doe had to let us campaign in relative peace. He could not abuse us or even kill us, not with observers from the United States and the international community swarming around. Still, there were consistent reports of soldiers using low-level intimidation tactics to steer the election. We just surged forward as best we could.

Election day, October 15, 1985, dawned hot, sunny, and dry. The voting lines outside Monrovia City Hall curved down the long driveway and staircase, curled up and down the sidewalk like twin snakes, and stretched nearly as far as the eye could see. General Doe arrived shortly after the polls opened at 8 a.m., and the crowds politely stood aside to let him be the first to mark his paper ballot. Afterward, he confidently reiterated his prediction that he would win. Seeing which way the wind was clearly blowing, few people disagreed.

In some ways the day was a victory for Liberia. For the most part, the elections went off peacefully, with hundreds of thousands of Liberians exercising—or trying to exercise—their democratic rights with patience and calm and in exceedingly high spirits. Old women walked from their villages in the scorching heat to stand in long lines at the polling places. My party's symbol at the time was the rooster, and I remember the crowds lining the road to cluck and flap their elbows as a sign of support. Long lines of people snaked at every polling place. Anyone who saw their enthusiasm, like me, could have no doubt that Liberians yearned for democracy.

But even as the day got under way, independent observers began to warn that something was seriously amiss. Poll watchers from our party and others began being barred from various voting places. Reports of various abuses flooded in: multiple voting and children casting ballots at the military bases, marked ballots disappearing.

By the end of the day our own LAP exit polls showed Jackson Doe having swept the presidential race with at least 60 percent and perhaps as much as 70 percent of the nationwide vote. However, it would take another two weeks for the hastily organized Special Elections Commission—appointed and controlled by Doe—to announce its "official" results. The commission went into seclusion for the counting of the ballots, ignoring the laws that dictated otherwise. While we waited, we took pictures of ballots being burned outside the city. During this time, General Doe stepped up his intimidation tactics, threatening to lash anyone who criticized him, while one of his justice ministers threatened to prosecute anyone publicly challenging the slow voting count. The count itself was conducted in secret, with so many reports of ballots being stolen, burned, replaced, or simply left blowing along the side of the road, that it was difficult to keep track. In the midst of the vote count, the *New York Times* ran an editorial saying it was clear that the elections had been neither free nor fair, despite the United States' warnings.

On October 29, the head of the Special Elections Commission, Emmett Harmon, announced the results: General Samuel Doe had scratched out a victory with a slim but decisive 50.9 percent of the presidential vote. In other words, he had miraculously garnered just enough votes to win the election outright but not so many votes that no one in the international community could even possibly appear to take the results seriously. Clever indeed.

Doe's running mate, Harry Moniba, was elected vice president, and their National Democratic Party of Liberia won 21 of the 26 Senate seats and 51 of the 64 seats in the House of Representatives. All in all, an impressive victory. And utterly, utterly false.

Among the five Senate seats Doe's party had not won was mine. I was the highest vote getter in the Senate. It was a bittersweet triumph at best. Independent observers agreed that Jackson Doe had, in reality, won a plurality of votes in the presidential election. Four days before the official announcement Jackson Doe delivered a powerful and eloquent "victory" speech to his supporters, which was later printed in the newspapers and widely read. The public knew exactly what was going on.

Together with the Unity Party, whose ticket I would run on in 1997 and 2005, and the Liberian Unification Party, our party mounted a challenge to the election results, sending formal complaints to the Special Elections Commission, having those complaints published in the newspapers, and, most critically, boycotting the seats we had won in the legislature. In the end, none of these challenges except the boycott amounted to much. The truth was that General Doe controlled the electoral commission and the army too. He claimed the election, named himself president, and declared the matter settled. He was now president-elect Samuel Doe, and there was nothing anyone could do to stop him, or so we thought.

As it turned out, not everyone agreed.

## CHAPTER 8

# THE ATTEMPTED COUP

IF IT is easy to pinpoint the best day of my life—inauguration day 2006—it's even easier to pinpoint the worst. That would be November 13, 1985, one day after an attempted coup d'état against the now formally, if fraudulently, elected President Samuel Doe. A coup attempt in which I had no significant part but for which I would pay a heavy price.

Although I had won my campaign for the Montserrado County Senate seat, I refused to take my position in the legislature as a protest against Doe's theft of the presidential election. President Doe was furious with my decision; for days he and his people pressured me to reconsider, to just accept the results and take the seat.

Even the U.S. State Department wanted me to take the seat, urging me and the others to "get on board." We were told to just trust the United States as far as Doe was concerned; everything would be fine if we just went along. Despite nearly universal condemnation of what had clearly been fraudulent elections, the Reagan administration granted Doe's theft of the presidency an almost unreserved seal of approval. Chester Crocker, the U.S. assistant secretary of state for African affairs, consistently and constantly applauded the elections, saying that their "shortcomings" should not obscure their "noteworthy, positive aspects" and downplaying any reports of irregularities by claiming there had been "injustices on all sides." It is hard to imagine that Crocker was not being deeply ironic when he praised Doe for claiming only 51 percent of the vote. It was, he said, "unheard of in the rest of Africa, where incumbent rulers normally claim victories of 95 to 100 percent."

Testifying before a skeptical U.S. Congress the December after the elections, Crocker called election day "a remarkable achievement. . . . The Liberian experience was not only important for Liberia but for all of West Africa." Going on, he said, "There is now the beginning, however imperfect, of a democratic experience that Liberia and its friends can use as a benchmark for future elections—one on which they want to build." As a result, he said, the United States would continue to stand by Doe: "To walk away would be irresponsible and clearly viewed as such elsewhere in Africa, where we are seen as having a unique responsibility for assisting Liberia. Abdication of that responsibility could provoke chaos and bloodshed, endangering the lives of nearly 5,000 Americans in the country."

Clearly the United States felt that my taking a seat in the Liberian Senate would lend the sheen of credibility to Doe's blatantly fraudulent elections. Which was precisely the reason I had to refuse. Today, however, I agonize, after the horrible aftermath of those elections, regarding our decision and wonder if we should have been more tolerant of the failed experiment in democracy.

Even though I would not take the seat, I did decide to remain in Liberia. Having tendered my resignation to Citibank, I had no immediate plans and no place more pressing to be than at home than with my mother, my son, and my country during what was clearly becoming an increasingly troubled time. The populace remained angry at the election fraud. General Doe grew more oppressive, and rumors of human rights abuses by soldiers swirled. Tension filled the air.

A few of us had been alerted that Thomas Quiwonkpa, who had been living in exile in the United States, had gathered some people and was planning to mount a coup attempt against Doe. When I heard about it, Quiwonkpa was said to be in Freetown, the base from which he planned to mount his attack. I was encouraged that some people, including Harry Greaves and Jim Holder, a popular basketball star who was the minister of commerce in Sawyer's government, were trying to get more information and were waiting for a message from some people in Freetown, including Henry Fahnbulleh. But for whatever reasons, I don't think they were ever able to put their hands on the message. Everything was unclear.

In the end, early on the morning of November 12, 1985, Thomas Quiwonkpa covertly entered Liberia through Sierra Leone at the head of a small core of heavily armed men and launched a coup against Doe. The people of Monrovia heard gunfire erupting in the streets early that morning. Then, before dawn, came an announcement over the radio, a voice declaring that patriotic forces under the command of General Quiwonkpa had gained control of the radio station, surrounded the Executive Mansion, and toppled the Doe regime.

"Our forces have completely surrounded the city," said a man who identified himself as Quiwonkpa.

The city erupted in celebration. In the six long years since Doe had taken power, his popularity and support had steadily eroded; his stealing of the election had put any lingering popular support into its grave.

Quiwonkpa, on the other hand, was highly regarded by many. One of the original group of soldiers who had carried out the coup of April 12, 1980, Quiwonkpa had become commander general of the army. Many saw him as a force for moderation in the Doe regime, eschewing the accumulation of personal wealth, fighting the growing corruption, and constantly pressuring the PRC to return the country to civilian governance, as it had promised. Quiwonkpa had finally broken with Doe in 1983 and fled to America. Word that he had returned from exile to topple Doe had many people dancing in the streets.

Some of those people came dancing into our yard, followed by Harry Greaves, who raced in in a pickup truck. He was rejoicing too, but he was also concerned.

"I am going to find Quiwonkpa," Harry said, "because the man is doing some silly things."

Harry said he had heard that Quiwonkpa was parading around the city and announcing victory when in fact he had yet to reach the mansion or to secure Doe himself.

"I'm going to go to find him," said Harry and left.

I admit it: I jumped up and down a few times myself, happy that the country seemed finally to have been liberated from Doe. Because of that shouting later on people would claim that we—myself and Harry and others—had not only known about the attempted coup but had masterminded it. That was simply not true.

However, being an active person, I did want to get on the streets to see what was going on. A friend of mine, Robert Phillips, and I got into a vehicle and told the driver to take us out into the city to see what was happening.

People were dancing and celebrating all around, but the celebration proved premature. Just as we were driving around to locate some friends, another announcement came over the radio. This was a different voice from the one that morning, with a very different message to deliver.

"I take this opportunity to inform the nation that the coup has failed," declared General Doe. "I am still the commander in chief of the armed forces of Liberia and head of state." He called on the people of Liberia to resist Quiwonkpa and his forces.

Word raced through the streets that Doe's people were now on the rampage, ready to wreak revenge on anyone. I turned to Robert and said, "If anybody sees us on this road now, we will surely be killed."

I knew a former police director by the name of Edwin Harmon who had recently passed away; the announcement of his death had just been made. I said to Robert that we should go to Harmon's house and visit his family as if we had set out that day to express sympathy. That way, if anybody found us on the street we could say we were coming from Harmon's house. Which is what we did.

We stayed at that home for about half an hour, then found our way home as quickly and invisibly as we could. Robert dropped me at my house, then went to his own place. Fortunately, we did not encounter any soldiers along the way. But my good luck would not last.

Later that day I was on the telephone talking to Jennie in New York when I heard the sound of a Jeep roaring into the yard. Soldiers, drunken and angry, began circling the house, shooting randomly. I knew immediately that they were Doe's men and that he was rounding up anyone he considered his opposition. I put the telephone down.

The soldiers were shooting wildly and randomly into the air. My mother, being the prayerful woman that she was, immediately began to pray. She got on her knees right there in the living room and began asking for protection. But I knew the soldiers were wild and drunk and irresponsible. If I didn't go out, they would storm the house and they wouldn't care that my mother was praying. They would shoot her without a second thought. The idea terrified me. So I took a deep breath and stepped outside.

The Jeep squealed to a halt.

"I know you came for me," I told the soldiers. "My mother is in the house. Don't harm her. You came for me, and I will go with you if you leave her alone."

They said, "Get in the Jeep."

So I got into the Jeep.

At that moment I was calm, very calm. I knew, of course, that the situation was serious, but I also knew I had to remain calm to have any chance at outwitting the soldiers. The soldiers demanded that I take them to the home of Jackson Doe. Although I knew where he lived, I told them I did not. Quiwonkpa had said on the radio that he had staged the coup not to amass power for himself but to install the right-fully elected president of Liberia, Jackson Doe. For that reason, I knew that Jackson Doe, if anything, was in even more serious trouble than I. I didn't want to help lead the soldiers to his door.

As it turned out, my gut-level reaction may have helped to save my own life. I have no doubt that had the soldiers taken me straight to Doe, I would have been killed immediately. But they had been told to bring in both "those troublemakers," and they didn't want to go back to their boss with the job only half done.

"I think Jackson Doe lives somewhere on this side," I told the soldiers, pointing in the direction away from the mansion.

"Come," they ordered, "we will find him."

We drove around town for a while, then headed toward Roberts Field, and while we drove I tried to figure out how I was going to get myself out of this mess. Just as we reached the gas station at the intersection that leads to Roberts Field, we saw another Jeep roaring toward us on the street. One of the soldiers signaled for our Jeep to pull over. By the swiftness with which our driver complied, I could tell the other man was a senior officer.

Standing up in the Jeep, he demanded, "Who do you have in your Jeep?"

"We have the prisoner Ellen Johnson Sirleaf," the leader of my little group replied. "She's got to find Jackson Doe for us. Then we're going to take the two of them to President Doe because they created all these problems for him."

But the officer shook his head. "This prisoner is not yours," he said. "We're coming from Shefflin, and we have instructions to kill them all there."

Camp Shefflin was the army's ragtag military post on the outside of Monrovia. The base, now renamed Camp Edward B. Kesselly in honor of the founder of the Unity Party, has recently been renovated and upgraded to provide more than a hundred clean and attractive housing units to our armed forces. But at the time I was taken there, it was a dump. It was also an exceedingly dangerous place to be taken as a prisoner.

Unfortunately, I had no say in the matter. "Turn her over to us," demanded the officer. The men in my Jeep leapt to comply.

Prodding me with their fingers, the soldiers ordered me from the Jeep. As I got out, one of them, a boy no more than twenty or twenty-one, made a move with his leg as if to kick me and knock me down. I saw it coming, though, and jumped out of the way. Then I turned and looked at him, straight into his eyes. I was trying to connect with him—as a person, as a human being, as a woman old enough to be his mother or his aunt. I wanted him to really see me; for soldiers in a war, it is so easy not to see.

The moment passed. The soldiers from the second Jeep grabbed me roughly by my arms and bundled me into the backseat of their vehicle, pressing in on either side of me and in front. The officer sat in front. Before we left, the top soldier from the group who had taken me from my mother's house turned toward the officer and said, "Okay. This is your prisoner. I'm putting her in your charge."

The officer said, "Yes." His name, which I will never forget, was Lieutenant Francis Harris.

We left the gas station and headed toward the military camp at Shefflin. That's when the terrorizing began. Taking a match from the pocket of his uniform, one of the soldiers struck it against the flap and held it near my hair. "I'm going to burn your hair off!" he said, laughing, inching the match ever closer to my hair. I concentrated on ignoring him, keeping myself calm and straight. Eventually the match burned down to his fingers and he blew it out. A few minutes after that, another soldier in the car noticed a small, gold ring I was wearing on my left hand. Grabbing my hand, he said, "You're going to die! Why are you going to die with a ring on?" Then he pulled the ring from my hand.

All the while this was happening, a kind of preternatural calm kept holding me steady. I knew that to have any hope at all I had to stay unpanicked and to try to outwit the men-children I was dealing with. The more they taunted and abused me, the more I tried to connect with them, to look them in the face and force them to recognize me.

By this point the sun was setting and the sky began to darken. Dusk was coming on as we reached the camp. The driver roared through the gates, then veered off the road and headed toward the beach. "This is where the grave site is," the soldiers said. "This is where we are going to kill you."

"Look at me," I said to them. "Think about your mother. How would you feel if someone did this to her?"

"We are going to take you there and bury you alive," one soldier jeered. The other soldiers were laughing and jeering and egging him on, but I just kept talking, kept trying to counter their aggression, kept trying to calm them down. Violence is a vicious, contagious disease, easily spread. To live I knew I had to inoculate at least one of them.

"Would you like it if someone did this to your mother?" I asked. "How would you feel? Think about her."

When we reached the beach, instead of stopping, the driver spun the vehicle in a big circle, kicking up sand. Then he turned it around and headed back toward the road.

"Okay," the soldiers said, laughing. "We won't kill you today. We'll take you to the jail and kill you later."

They drove back toward the front of the camp, to a building no bigger than the width of two cells, which is what it held. One of the cells—small, dank, lightless—was empty; I was hustled inside and the bars were slammed shut. The other cell held a crush of men, who were terrified.

"Give me the laces from your shoes!" one of the soldiers demanded of me. When I complied he snatched them from my hand, walked to the other cell, and used the laces to tie the wrists of two of the men. The men—I couldn't tell how many there were—were crying out now. One man had recognized me and called my name.

"Mama Sirleaf!" he called. "I know you, I know your mother! Help us! Try to save us!"

The terror in his voice, in all their voices, was palpable and shattering. They knew what lay ahead.

"How can I save you when I am here with you?" I called back. "How can I save you when I cannot save myself?"

My cell was no bigger than a closet and reeked of urine. In a corner at the back was a pile of sand meant to be used as a toilet, with a small alcove off to the side serving the same, but more extensive, intent.

For the next half hour or so I sat in the gathering darkness on the cold concrete floor, talking across the gap between the cells to the men, some of whom were beside themselves with fear. "What will happen?" some kept asking. "Why are they doing this? We are innocent! We did nothing. Help us, save us, please." I did not know what to say, just tried to offer comfort and hope as best I could.

Suddenly a group of soldiers came swarming into the building and opened the door to the cell where the men were being held. "Come!" they ordered and marshaled the men out.

The men were crying and begging, pleading for their lives. The one who had recognized me called my name over and over as he was dragged away. "Please!" he cried. "I know Ma Martha," which is what my mother was called by many of her former students. "Please, beg them for us! Don't let them kill us!"

All I could do was watch them take him and the other men away. I was left alone in the locked cell and in the building.

Ten or so minutes later, I heard gunshots in the distance. Those poor men were never heard from again. It was a terrible thing. I felt sickened and terrible but still not frightened. I knew the situation was very uncertain, that it could go any way.

Killing has a terrible way of exciting some people who have become accustomed to it. It revs the men up. That's what happened that night. After the executions the soldiers came ambling back toward the building where I was being held, high-spirited and feeling mean. Singly and in groups they would wander in to taunt me, to curse me and call me names, to make obscene gestures and point their fingers and jeer.

"Who are you?" they cried, banging on the bars with the butts of their guns. "Who are you to trouble our Pape?"

"You old, dry, red, funky woman!"

"Coming here to bother our Pape! Who you think you are?"

Obviously, sleep was the furthest thing from my mind that night. I didn't know what might happen from one moment to the next. The soldiers had taken those poor men out and shot them without a moment's hesitation, and they could just as easily end my life. On the other hand, they hadn't. Something was holding them back: the absence of a clear go-ahead from Doe, the understanding that I had international support, perhaps the fear that I had juju of some kind that would protect me in the end. Whatever it was, it was tenuous and I knew it. I prayed and breathed and tried to think.

Sometime close to midnight a soldier came into the jail and stood before the bar for a moment, watching me silently. I waited to see what he would do: curse me, fling lit matches my way, bang the bars to make me flinch. But he did none of those things. Instead, after a moment, he said, "I'm going to fuck you." Then he stepped forward and started to open the gate.

Rising to my feet, I steeled myself, although for what, I cannot say. What woman really knows what she would do in that terrible, most-dreaded situation? Would I have fought? I think so. Would I have tried, above all, to retain my life, to stay alive? I think that too.

But just as the soldier was opening the gate, a voice behind him said, "As you were."

The person must have been of higher rank, because the soldier immediately dropped his hands to his sides.

"Retreat," ordered the voice. The soldier closed the gate, locked it, and moved away.

So, one terrible fate avoided. But who knew what was coming next? I didn't breathe any easier yet. The soldier who had threatened to rape me went out the door, and the one who had stopped him moved forward, closer to the bars. He was slightly older than the young boys who had been taunting me, mid- to late twenties or so, medium build and angular, with beautiful dark skin and a very serious face.

"They said you are Gola," he said to me.

"Yes," I said. "My father is Gola."

"Say something in Gola," he ordered.

Now, I was far from fluent in Gola. My father had spoken it, of course, but like many Liberian parents trying to pull themselves up from poverty and the underclass, he had not focused on teaching his children the language of his roots. He had, however, always insisted we visit his village and his mother. And from her I had learned a few words. So I said, "Ma keye," which means "hello." And then, in Gola, I thanked him for what he had just done for me.

"Where your pa town?" he asked.

"Julejuah," I said. "My grandmother lived there, my pa lived there. I live there sometimes."

The soldier looked at me for what seemed like forever. Then he nodded. "Okay," he said. "I will stay here all night. Nobody will trouble you." He stayed, and no one did.

A few hours later, near 2 a.m., according to my watch, a soldier brought in a young woman and pushed her into the cell with me. She was a young Gio woman, no more than twenty, and she was terrified. She was totally naked and crying hysterically. I could guess what had happened to her, and I tried to comfort her as best I could. There were by that time a few soldiers milling around outside the gate in front of the one who was protecting me. I tried speaking to them, asking them to think of their mothers and their sisters, asking them to at least give the young woman something to cover herself with. Finally someone went and came back with a *lappa*, or wrapped skirt, for her to put around herself. This calmed her a little, and the two of us spent the rest of the night sitting together on the floor of the cell, wondering what the dawn would bring.

Early the next morning they came for me. It was the same Lieutenant Harris who had brought me into the camp. "You need to go speak to the commander," he said, pulling me from the cell.

He and another soldier marched me from the jail building to another building deeper into the camp, clearly the headquarters. We entered a room where five or six soldiers, all of them high-ranking, sat waiting around a table. One of them introduced himself. His name was General Moses Wright, and he was commander of the Camp Shefflin brigade.

"Why are you causing all this trouble?" General Wright asked me. His tone was calm, nonthreatening.

"I'm not causing any trouble," I responded. "I have nothing to do with whatever has happened here."

"Why don't you want to work for President Doe?" he asked. "Just go and take your seat in the legislature."

That, then, was the crux of it. My refusal to take the Senate seat in protest of the fraudulent elections was causing President Doe no end of international embarrassment. As time went on, more and more of

the eighteen non-NDPL candidates who had won seats in the Legislature would end their private boycotts and assume their positions, but
I could and would not.

So I said, "I'm sorry, but I can't do that. "

General Wright and Lieutenant Harris tried to urge me on. "Just
go and work with him, and things will be all right," they said. "Take
your seat."

Still I refused. I told them my party had decided that the elections were fraudulent and that I could not be a party to that fraud. "If
I accept, I will lose all credibility with all of the partisans who supported me all over this country."

They said, "Okay. That's your choice."

The men conferred among themselves for a few minutes. Then
General Wright turned and told me to go back to my cell and prepare
to do some work around the camp, since I would be staying there for
a while. I was taken back to jail and ordered, along with the young
woman they had brought in the night before, to go out and cook a
meal for the soldiers.

They took us out to the field in the middle of the barracks where the
cooking was done. The first order of business was to clean the pans,
which the soldiers, cooking for themselves, had left in dirty heaps upon
the ground. Someone brought us a bunch of coconut shells that we could
use to scrub the pans because they had no soap for washing. The young
woman and I got busy scrubbing the pans and making the fire, while
the soldiers made themselves comfortable on the ground to watch.

After a few minutes of us scrubbing, someone said, "You need water.
There is a pump over there. Go to the pump."

It was clear the soldier was talking to me; whatever the poor young
woman had done to be locked up, it was nothing compared to my
crime. The soldiers wanted to humiliate me, to put me in my place. I
knew I would have to bear whatever they came up with.

By this time we had an audience: the families of many of the soldiers at the camp had come out of their huts and lined up along the
field to see what was happening. I picked up a bucket lying on the

ground, walked to the pump a few yards away, and began filling the bucket. It was an old pump and slow, and after a few minutes the soldiers became impatient. "Why are you taking so long?" they called. "We want to eat!"

"I am waiting for the water to pour out of the pump," I responded. "It's coming slowly."

"Well, then, talk to the pump!" one of the soldiers cried. The others, laughing, took up the call. "Talk to the pump!"

So I talked to the pump. "Pump," I said, "can you please pour the water fast because the soldiers want to eat. Hurry, pump."

I kept talking to the pump, ignoring the laughter and jeering, until the bucket filled. Then I carried it back toward the cooking area and began to prepare the rice. But before I could get much further Lieutenant Harris roared up in his Jeep.

"I told the general I came for my prisoner," he said to me. "I'm now taking you to President Doe. You have to face him."

Suddenly, for the first time in the whole long and ridiculous ordeal, I was afraid. The sense of calm that had carried me through so many dangerous situations, the inexplicable but unshakable belief that things would work out for the best, suddenly deserted me, and into its place crept terror. I knew that if I were taken before Doe, the order would be immediate execution. I would no longer be dealing with young soldiers whose sympathy I could hope to penetrate. I would be dealing with the ultimate battle-hardened and entrenched general. He would tolerate me no longer.

They put me back in the Jeep, and we drove off. From the window I watched the young woman continue her cooking. I never learned her name, and I never saw her again.

This time the drive was quiet. There was no taunting or jeering; the two soldiers on either side of me and Lieutenant Harris in front were all silent and completely straight-faced. From Shefflin to the Executive Mansion is perhaps a thirty-minute drive on the best of days. We made it in silence, me praying all the way.

When we reached the mansion, Lieutenant Harris told the driver to drive to the back. There we found a man whom I would later learn was Colonel Edward Smith, the man in charge of the Executive Mansion Guard.

Smith was a strong, determined-looking man, a soldier's soldier. But there was clearly something else at work beneath the uniform, because he looked at me a moment, then asked, "Why are you making all this trouble in this country?"

I repeated what I had been saying all the day before: that I had no intention of making trouble and had nothing to do with the coup but that I could not take my Senate seat.

Smith seemed to sigh, then said, "We are all one people here. Why are we fighting each other this way?" Without waiting for an answer, he turned to Harris.

"Don't take her to the president," Smith said. "You know what will happen if you do."

Harris knew all too well what would happen, and so did I. Standing there, looking at him, I waited to see what my fate would be.

"Take her to the BTC," said Colonel Smith. "Tell them to put her in prison there."

As the Jeep started to drive off, we heard a voice. It was Colonel Smith calling out, "Stop." The driver stopped. Smith walked up and addressed Lieutenant Harris. "Tell them I said not to harm her. Just put her in prison and don't do anything to her. On my command."

So I lived. The BTC, Barclay Training Center, was no picnic, but I had been imprisoned before and weathered it. I knew that as long as God kept me from the hands of the killers, I could survive the discomfort and even pain of prison.

In the many years since that day I would meet again and speak to these men—Colonel Smith, General Wright, even Lieutenant Harris. And I would thank them, for each, in one way or another, saved my life. It's a curious thing. I believe that deep inside they were good people, even though they worked for Doe and were part of his tribe and thus inclined to protect him, to go along. They would carry out his instructions, and they did—but only so far. They were not prepared to go the extra mile to do something that they knew inside was not right.

I know that had fate led me that morning into the hands of some of Doe's close bodyguards instead of Colonel Smith, they would not have asked any questions. They would have killed me on the spot, as they killed so many others in those brutal days following the coup.

To this day, no one knows how many hundreds, or perhaps thousands, of people were killed in the days following the attempted coup. We do know that Doe focused the brunt of his retaliation on the Gio and Mano of Nimba County, who paid a heavy price for being from the same ethnic group as Quiwonkpa. We know too that Quiwonkpa himself was shot and his body put on public display at the Barclay Training Center. Doe had imposed a 6 p.m. curfew and threatened to shoot anyone, including foreign nationals, found on the street as little as one minute after that time.

What saved my life that time and all the terrible times before? What stayed Doe's hand? I have often wondered that myself, because the truth is that with all I went through, I should be dead. People who were seen as being a far lesser problem to the Doe administration lost their lives during his brutal and oppressive reign. There were so many times Doe could have killed me, could simply have given the order to one of his wild and fearsome bodyguards, men who would have relished the opportunity.

There was, I think, an element of fear on the part of Doe and his closest advisers, a fear that I represented something bigger than myself. They remembered what had happened the last time I'd been imprisoned, the outcry both internally and externally, the incredible pressure from the highest levels of the U.S. government. That restrained them somewhat, I think, caused men not accustomed to pausing before acting to do so and ask themselves first what the consequences might be.

Somewhere during that time I learned that Doe had also arrested and imprisoned my son James, who had absolutely nothing to do with anything even remotely political. Naturally I was worried. Fortunately, though, he was never hurt and was released a few months before I.

On January 6, 1986, Samuel K. Doe was sworn in as Liberia's president. In his inaugural speech he appealed for reconciliation across the land and expressed hope for continued friendship with the United States. He also released eighteen people who had been arrested after the coup, including my son James. He did not release me.

Later that month Jackson Doe (who had been located and detained without my help) and I were both charged with sedition, a capital crime. Once again, an international outcry arose and put pressure on Doe to stand down from his oppressive tactics. Many in Washington had grown weary of supporting a regime so blatantly fraudulent and increasingly brutal. Doe had used the Quiwonkpa coup as a reason to crack down even more on the population, waging, in particular, a bloody retributive campaign of terror against the Gio and Mano of Nimba County. The BBC reported that as many as 2,000 Liberians may have lost their lives in the brutal and terrifying first days and weeks after the attempted coup; some sources place the estimate at closer to 3,000. No one will ever know for sure.

In February, Jackson Doe and several others who had also been arrested in the wake of the coup were released from jail. I, along with some others, remained imprisoned for five more months, until July 1986, when the international and domestic pressure finally proved too much for Doe to resist. Once again Jennie and others had rallied people all over the world to our cause; Jennie had even been interviewed on several major American news programs. Doe gave in. In a radio speech to the country he announced our "pardons" in the name, he said, of national reconciliation and to promote "peace and stability" in the country. Finally, nine months after the confrontation at my mother's house, we were free.

The morning I and the other remaining political prisoners were granted amnesty, soldiers arrived at Central Prison, where we were being held. In the prison yard one of the solders said I should get into the front seat of the truck they had brought while the male prisoners would ride in the open back. I said no. Since I had shared the same cell and the same sentence as the other prisoners, I would ride in the back with them.

As I was climbing into the back of the truck, a soldier walked up and put his hand on my arm to get my attention. I did not recognize him at first.

"Yes?" I asked.

He reached into the front pocket of his uniform shirt and brought out a ring. My ring, the one that had been taken from me the day after the attempted coup.

"Oh!" I said, astonished that he had kept the ring that long and that he was now returning it to me. "Lieutenant Harris!"

"Colonel Harris, please," he said, handing me the ring. Whether his promotion had come because of, or in spite of, his handling of my arrest and imprisonment I never knew. Later, after the war, I tried to look him up to thank him for his help, but he had died. So many, many people died during those wars.

I got on the back of the truck with the others and as it pulled out into the street, the crowds materialized and lined the sidewalks, cheering us on. Once again, we had become folk heroes to a people tired of governmental oppression and wrongdoing. I couldn't believe the crowds. They followed the truck all the way from Central Prison to the Unity Party headquarters, where we stopped and began an impromptu rally. Jackson Doe, who had been released ahead of me but would soon be imprisoned, yet again, stood at my side.

Someone brought a small table out into the street and I climbed on top of it, raising my fist into a power salute. The crowd erupted with noise.

"Freedom!" I cried.

"Freedom!" the crowd roared back.

"We will overcome! Tyranny will not prevail!"

*"Aluta continua!"* the crowd roared. *The struggle continues.*

That was a wonderful day.

# CHAPTER 9

# ESCAPE

_____

A FEW days after I was released from prison, President Doe sent for me. Not having much choice in the matter, I heeded the call. We sat together on the porch of the Executive Mansion, and he spoke to me as if we were old friends. It was all very strange.

"I do not understand why you can't take the seat," he said, sounding more weary than upset. "We want you to take the seat."

By this time some members of my party, led by our chairman, Tuan Wreh, had dropped their opposition to Doe's fraudulent election and agreed to take their seats in the legislature. Wreh withdrew our party's complaint to the Special Elections Commission and issued a statement saying the party had decided to take the seats it had won "consistent with the spirit of promoting genuine national reconstruction, unity, peace and understanding."

As a result the party's executive committee—dismayed by these actions—threatened to suspend Wreh and the others unless they reconsidered. Instead Wreh and several others resigned from the party and vowed to take their elected seats.

Doe wanted to know why I could not do the same: forgot the protest and go ahead and take my seat. In response, I said that I was sorry but I simply could not do as he wished.

"Mr. President," I said. "I'm sorry, but I've told you I cannot take my seat because the election was fraudulent."

I told Doe that if I took the seat it would send a very large and disturbing signal to all Liberians: that I was granting the election my seal of approval as legitimate. It would signal, essentially, my willingness to be bought. That was a message I refused to send, regardless of the consequences.

"Mr. President," I told him, using the term lightly, "I have to continue to refuse."

I wasn't sure what his reaction might be, whether he would get angry or deny it or what. I braced myself for anything, but Doe simply shrugged and raised his hands.

"The election is in the past," he said. "It's done. What can I do about all that now? We must put the past behind us so the country can move forward. We must let all these things be forgotten and move ahead."

This was, at least most visibly, Doe's tactic during those months after his fraudulent election: entice, divide, and conquer in the end. After rounding up every possible political opposition leader in the days immediately following the failed coup, he had gradually released most of us—a few here, a few there, no doubt with the hope of currying favor. Likewise, in the first days after the election rumors swirled that Doe would offer Gabriel Kpolleh the position of minister of education, an offer aimed at undermining Kpolleh's credibility and dividing the ranks. Later, he actually did offer the post to Mary Brown Sherman, the ousted president of the University of Liberia. She refused.

What General Doe failed to see, like President Tolbert before him, was that a people and a society that have been radicalized will not be turned back. Progress may be slowed by oppression, but it will not be stopped.

That this was so was clear from all the agitation taking place in the country. Journalists were constantly threatened and harassed by both soldiers and judicial officials, yet they continued to publish independent newspapers as best they could. In March, schoolteachers had gone on strike to protest a four-month delay in their salaries being paid by the government. Schoolchildren staged a series of protests in support of their teachers. When police responded to the protestors with force, riots broke out.

While I was still in prison, my party had joined with the Liberian Unification Party (LUP) and the Unity Party to form a grand coalition, whose goal was to struggle for genuine democracy in Liberia and resist Doe's clearly signaled intention of creating, yet again in Liberia, a one-party state. Our presidential candidate, Jackson Doe, was chosen as vice chairman along with Dr. Edward Kesselly of the Unity Party. Gabriel Kpoleh of the LUP was picked as chair. The coalition called for Doe to step down as president, citing not only the rigged elections but the increasingly political instability of the country and the dire economic straits.

By this point, Liberia's economy was in a near tailspin, with upward of $900 million in outstanding foreign debts. Because of its record of nonpayment, the country had been cut off from further loans from the IMF, the World Bank, and the United States.

In April, the Reagan administration sent Chester Crocker to the country to try to negotiate some compromise between Doe and the grand coalition. But by then the opposition was in no mood to negotiate. Speaking for the coalition, Byron Tarr, LAP secretary general, said, "It's too late for a compromise. We have tried to work with Doe in the past, but he has made it clear that he is not genuine."

He added, "If it was Crocker's intention to seek a compromise, it will fail."

Doe responded by holding celebrations to mark the sixth anniversary of the April 12 coup, by opening a multimillion-dollar sports stadium, built by the Chinese, in his name and a million-dollar market, built with Liberian tax dollars, in the name of his wife, Nancy Doe.

Nonetheless, at our meeting together I told General Doe that it was not too late to correct his actions. Although I knew only too well that he would not go so far as to overturn the fraudulent results, I suggested he might signal to the country and the world that he was willing to correct the mistakes of the past and move forward on a different note: he could remove all the people sitting in the legislature who had not been honestly elected. He could boot them out and send them somewhere, so that the people who had really won could take their seats. I went on to list a few other things he could do, both to begin putting the country's increasingly disastrous economic house in order and to begin seriously moving the nation toward a true democracy.

Doe nodded as I spoke. Then he said, "Okay. Everything you say, go and write it up so I can read it and we will see what I can do."

So I went home and wrote a memo outlining all my suggestions for improvement, then sent it in to him at the Executive Mansion. Nothing happened; I heard not a word about it until months later, after so much had changed. Shad Tubman, the oldest son of the former president and a former senator, came to me one day and said, "Do you remember that memo you sent to Doe?"

I said that I did. By this time Mr. Tubman was an ex-senator from the Tolbert administration but one who was popular because of his father.

"Do you know why it was never taken up?" he asked. "Because when you wrote the memo you didn't address him as President Doe. He felt insulted."

I just shook my head. What can one do with a leader like that?

At any rate, I had no time to worry about whether or not Doe would heed my advice. After the meeting at the Executive Mansion, it became all too clear to me that I was under surveillance. I was being followed, tailed, every single day as I went about my business, men in unmarked cars following me around, trying to remain unseen. Fortunately I had many good people around me, driving me and carrying me here and there. Sometimes we would try to lose them, accelerating quickly in the car and turning off down some side street before the people following me could catch up. Sometimes I would sneak into my car and lie on the floor so they wouldn't know who was in the vehicle. Following the advice of friends and my own inner sense of self-preservation, I began changing my movements, not keeping a set schedule, sleeping in a different place every night.

People began calling me, warning me to be careful, warning me that the rumblings about me in the corridors of power were growing louder by the day. In July, during a trip to the town of Buchanan, I was surrounded by well-wishers as I walked down the street. The next day a newspaper in Monrovia reported that I had staged an illegal rally and a local senator had called for my arrest. Fortunately I was then safely back in Monrovia and nothing happened, but it was a troubling sign.

One evening a man from Doe's tribe, a former ambassador named Peter Johnson, came to me and said, "You know I come from the inner circle, so listen to me when I speak. You better leave the country. You better leave soon because they are planning to get you."

I looked at him. Needless to say, this would be much easier said than done. I couldn't just start asking around about ways to slip out of Monrovia. The Doe government still held my passport, had already denied my application for an exit visa, and had warned me and the other released political prisoners to remain in the country. Any indication that I intended to do otherwise would be quickly noted and acted upon.

Still, Mr. Johnson was insistent. "I don't know how you will leave," he said. "But you better find a way."

(Incidentally, it was the same Peter Johnson who, when young partisans had joined street demonstrations to protest the electoral fraud, had persuaded me to intervene to call them off, so that I would not have their blood on my hands. We did not anticipate that, in fact, thousands would subsequently die in part because of this acquiescence.)

It is shocking to be confronted by such an announcement, even if you have suspected all along. In my heart I knew he was telling the truth. I had been a thorn in Doe's paw for a long, long time. It was clear by now that he was reaching the outer limits of his tolerance. At some point he would reach it, and soon he would cross the line. My international reputation and luck had protected me in the past, but at some point I knew Doe would say to himself, "So what if the United States doesn't like it? Once she's dead, she's dead."

That was not a point I wanted to reach.

In early August, Jackson Doe was arrested, along with Gabriel Kpolleh of the Liberian Unification Party and Dr. Edward B. Kesselly of the Unity Party. All three were sent to Belle Yalla prison. The arrests sparked a mass demonstration by the women of Liberia, and I took part in the march, despite a warning not to do so from Keikura B.

Kpoto, chairman of the NDPL. "Something might happen to you," he said, "and no one would be responsible."

Kpoto later denied that his words had been a threat, but I took them as such. Two days later, my home was broken into by some men in civilian clothes. They smashed up everything and told neighbors they were looking for me.

As much as I wanted to stay in Liberia, I wanted even more to stay alive. It was time to go.

We made all the arrangements through my friend Clave. She connected with a man who had extensive connections throughout the country, a man who had married the widow of Stephen Tolbert but had also established a connection with General Doe. His name was Gabriel Doe—no relation to the general but from the same part of the country. It was Gabriel Doe who arranged a private plane to whisk me from Liberia.

We decided I would leave on my son's wedding day. It would be the perfect diversion—the wedding had been publicly announced, the invitations sent and responded to. Everyone would assume that I would be busy preparing for it and certainly present that day; no one would ever dream that I might miss such an event. We hoped the expectations would be enough to convince my pursuers they could ease off for one day and give me some space.

That night I slept at Clave's house. At three in the morning we rose and dressed: me in jeans, a bulky red jacket, a cap pulled down low over my head, and glasses to hide my face. By 4 a.m. we were sneaking out of Clave's apartment, down the stairs of her building, and into a car waiting outside. The city was still sheathed in darkness, with few people on the street. Nonetheless, I lay on the floor of the car until we had cleared the outskirts of town; then I sat up.

It would have been much too dangerous to try to take a plane out of Roberts Field. So instead we drove to Buchanan, a town about ninety miles southeast of Monrovia, where Gabriel Doe had a small plane waiting on the airstrip. By dawn the plane was rising over the countryside. Looking down over the fierce green carpet of the Liberian landscape, I felt a pang of sadness. I didn't know when I might see my homeland again.

There was some confusion in the air; the pilot got either lost or confused about where he was supposed to land or he simply overshot the mark, because we ended up at the Tabou border, where we had not been expected. Having not been cleared to land, we all had some concern about whether or not we would be allowed to enter the country. Needless to say, we couldn't go back, and the plane was too small and too low on fuel to go on.

But I knew that the president of Côte d'Ivoire, Félix Houphouët-Boigny, held hard feelings against Samuel Doe. Houphouët-Boigny's daughter had been married to A. B. Tolbert, a son of the late president. A. B. Tolbert had taken refuge in the French Embassy after the 1980 coup, only to find his desperate pleas for asylum made null and void when a gang of unruly soldiers burst into the embassy and snatched him up. Houphouët-Boigny's demands that Tolbert be released were rejected by Doe, and when Tolbert was later killed Houphouët-Boigny was furious. All of his hard history, I knew, would mean sympathy for my own plight.

We landed, and the people with me quickly went out to find the local prefect, a man who is still alive today. After the situation was explained to him, the prefect not only granted us asylum but took me to his room, where he and his wife actually gave up their bed space for me until we could make arrangements to go on to Abidjan. These were incredible, warm, and giving people, and I am eternally grateful to them. We stayed in that place overnight, then, the following morning, took another plane to Abidjan. Even though I was no longer in Liberia, I was still anxious, as were those traveling with me. For all we knew, someone had paid the pilot off. For all we knew, we were being flown not to Abidjan but back to Monrovia. Staring from the window at the West African landscape below, I said a silent prayer that we were headed toward freedom and not away from it.

And we were. Abidjan, when we landed, was its usual bustling, glorious, French-African self. The city offered more security, more anonymity, and, most important, a path to New York. I spent another day in hiding in that city, contacting a friend from Citibank, who quickly sent a plane ticket. The next day I caught a flight to New York. It was only when the jet cleared the West African skies that I could really breathe a sigh of relief.

By this time someone had called Jennie in New York to deliver the agreed-upon message: "The package is on the way."

That was the good news. The bad news, the really terrible news, was that by this time the Doe people back in Monrovia knew that something was amiss. When I had failed to show up for the wedding as expected, they had immediately begun to search for me.

They had gone to my home, to Clave's home and the homes of other friends, to the home of my son. We do not know for certain, but we think they checked every place they could think of until somehow they found out a plane had taken off from Buchanan that morning.

When Doe's security forces reached the airport at Buchanan, they encountered an immigration officer. They accused him of being a party to my leaving, of having allowed me to take off without alerting them, of having even gone so far as to stamp my passport. None of this was true. The man had had nothing to do with it. I didn't have a stamp on my passport when I fled Liberia. I didn't even have my current passport! Instead, we had taken an old passport and refaced it to get me into the United States. I planned, in the face of any resistance, to appeal for political asylum and to once again call upon my friends and contacts in Washington. But this was the pre-9/11 America; gaining entry to the United States was not the obstacle course it is today. Immigration was far less strict, and I managed to enter the country with no problems.

Still, the fact that I left unaided by anyone in Liberian immigration did not matter to Doe's security forces. They had to find someone to blame, and, horribly, they blamed the immigration officer. Terrified of his coming fate, he committed suicide.

To this day, the man's family blames me for his death. I am sorry for that. Obviously, I felt terrible that a man should die on my behalf, even though I had nothing to do with it. When I returned to Liberia years later, I went to Buchanan to meet this man's family and to express my condolences. I visited their church and I tried to help them when I could. Every now and then, when the family has a rough time or they need something, they bring the story back. Our father died because of you, they say.

At the time, I knew nothing of any of this. Landing in New York, I felt primarily relief, mixed with worry about my son still in Monrovia and regret at having had to miss his wedding. For safety reasons, he had been kept completely in the dark about my plans. He was as mystified as anyone when I didn't show up at his wedding. "Where's my mother?" he kept asking people. It wasn't until I was safely in the air that Clave could pull him aside and whisper the truth.

I worried for James and his new family, but, fortunately, no harm came to them. I can only guess that by that point President Doe knew there would be no point in it.

Fortunately, although he was jailed for some period of time, no serious harm came to Gabriel Doe. In his case, I am certain that had the truth of his involvement become known, he would have been killed. President Doe would not have let such an action pass with no response. But Gabriel Doe was a well-known businessman who had worked closely with Doe and his government. As such he was above suspicion, and it wasn't until many years later that the truth came out. I never told anyone outside the small circle who already knew, because I never wanted to put him at risk. At this stage I don't think he's at risk anymore. Liberians have had to learn to let go of many things in the past in order to move forward. After all, Prince Johnson, the rebel leader who killed President Doe, is now a senator.

There you go. Strange country, some people might say.

When I reached New York, my sister Jennie said, "Thank God you're out of there." She and her family, like many others, had left Liberia in shock after witnessing the terrors and devastation that had followed the April 12 coup.

Like many Liberians, such as myself, Jennie knew there were distinctions being made within our society. But she had had no idea of the depth of those divisions or the anger than ran through them until the day of the coup.

"Growing up, we looked at each other as Liberians," she says. "There was never this feeling of hostility that we could see, and so when the coup took place in 1980 we were appalled and shocked. This for us was home, and everybody seemed to love everybody else. For this terrible violence, then, to happen was devastating."

Jennie is right when she says that most of us were taken completely by surprise. Over the years before 1980, we had known there was a cleavage there. We knew there was resentment, and by the late 1970s we knew it had been politicized and radicalized. But people generally lived together and worked together and prayed together. Nobody really saw that the division ran so deep it would spill over into hatred—violent, boiling hatred. That's what surprised everybody—the extent of the violence. The inhumanity of it.

Like many people, Jennie was heartbroken. She told me she had no intention of returning to Liberia, and she could not imagine why I, of all people, would want to do so.

Strangely, though, I never reached the point of not wanting to come back home to Liberia, nor of believing that some day I would. When push came to shove, I knew I had to leave when I did, but I was always, always, always looking for the opportunity to return.

CHAPTER 10

# EQUATOR BANK AND THE CHARLES TAYLOR WAR

ONCE AGAIN I found myself living and working in the United States. After some time in New York with Jennie, I moved to Hartford and took an executive position as vice president and director of Equator Bank.

Equator Bank was part of the giant Hongkong and Shanghai Banking Corporation, which specializes in granting lines of credit to so-called high-risk countries, many of them in Africa. The chairman of the bank, Niles Helmboldt, was one of the people who had gone to bat for me during my imprisonment, testifying before Congress and rallying international support. He leaned heavily upon Deborah Harding and Steve Cashin, and Steve has remained until this day a very strong partner in everything that I do. Once I was safely in the United States, I contacted him to thank him for his help, and he asked if I would consider coming to work for the bank.

The position dovetailed perfectly with my skills and abilities and allowed me to continue my overarching goal of helping Africa continue its steady development. The position also afforded me the opportunity to travel to Shanghai and to meet businessmen and officials in that burgeoning country, connections I would later be able to use to benefit Liberia.

Eventually, I was transferred to Equator's Washington office, and I returned to living in Alexandria. Life in America is less strenuous, in many ways, than life in Africa; the daily basics of survival in suburban

Alexandria were so much more easily met than in struggling Monrovia. What's more, most of my family, save my oldest son, was in America, along with many, many acquaintances and friends. My work was there, and my career was thriving. It would not have been difficult to let the troubles of my homeland drift gently to the horizons of my mind.

During this time my mother continued to be a major force in my life. In some ways she had grown increasingly attached to me as the years passed, as I was the only one of her children who did not have a spouse and young children at home and so could give her more attention than any of the other siblings. She was always such a big part of my life that my siblings sometimes joked that she showed an extra preference for me, but perhaps mostly it was that as the years passed and our oldest brother, Charles, died and Jennie and Carney both settled down to raise their families, I was alone for the most part, my children grown, my career moving me from job to job and country to country. My mother was with me much of the time: she lived with me in Nairobi, bringing a welcome measure of family life and family continuity into those years. In fact, I believe that having my mother nearby helped to stabilize me and make me content in a way that I didn't see the need to marry again in order to have companionship. My mother was my companion, and we got along wonderfully.

When I moved back to the United States, she came along and lived with me again. She liked Virginia. We went to church together and her friends visited, and she was happy and content. But as time went on, we knew she was tiring. There were times when she would say, "I want to go home now. I want to rest." We knew what she meant, so we'd protest. "No, no," we said. "You still have many years to go."

One evening my youngest son, Adamah, who is a doctor, was visiting us at home in Alexandria. We all had dinner together, then my mother said she wanted to go to bed. She went into her room and went to sleep. After a while Adamah and I heard the woman whom I had hired to help me with domestic duties and my mother's care crying out.

"Come quickly!" she called. "Something has happened to your mother."

*(Right)*
Ellen's boys: Charles,
*left*; Rob, *middle*; Jes,
*top right*; and Adamah,
*bottom right*.

*(Below)*
Ellen in her early days in
Monrovia at the beach.

*(Opposite, top)* Getting the boys off to school each morning:
*from left to right*: Rob, Adamah, Jes, and Charles.

*(Opposite, bottom)* Ellen at home in McLean, Virginia.

*(Above)* Ellen and Rob in Paris.

*(Above)* Rally after release from jail in Monrovia.

*(Opposite, top)* Unity Party rally during Ellen's first run for presidency.

*(Opposite, bottom)* Free at last from jail in Liberia.

*(Above)* Ellen's mother, Martha Johnson *(left)*, and a friend
hugging Ellen after her release from jail in Liberia.

*(Opposite, top)* Ellen dancing with Unity Party women at a party rally.

*(Opposite, bottom)* Protesting in New York.

*(Above)* Ellen speaking at a mosque in Monrovia.
*(Below)* With Nelson Mandela.

My son got up and rushed into the bedroom. But something strange happened to me. I don't know if this happens to other people in times such as that one, but my body suddenly turned on me and I had to go to the bathroom, intensely. It felt as if my whole inside were opening up and being ripped away. By the time I got out of there to go to my mother's room, it was over. My son turned to me as I entered and said, "She's dead."

He had called for an ambulance, which arrived shortly thereafter and took her to the hospital, where she was pronounced dead. We gave her a beautiful and fitting funeral, but we were not able to take her home at the time. She is buried in a cemetery in Alexandria, and in days past, whenever I went to the United States I made sure to visit her grave. We took all of the grandchildren there. She was very close to all of them who were alive then, and they miss her. As do I. But now, as president, it is not so easy to get to the grave, so I hope that someday soon I can bring her body home.

In the meantime it was clear that troubles were increasing at home. It was a highly frustrating time for those of us in exile, as we tried with only limited success to get the United States and the world to focus on the continuing struggle for democracy in Liberia.

In January 1987, George Shultz, the U.S. secretary of state, flew to Africa for a six-country tour that included, incredibly to many of us, Liberia. Shultz not only met with Doe, he came away from the meeting so impressed that afterward he declared that there had been "positive" and "genuine progress" toward democracy in Liberia. Shultz said the country had a free press—despite the fact that Doe had closed down several newspapers and others operated under careful and deliberate self-censorship for fear of retaliation. Shultz said there were no political prisoners, despite the fact that Doe was very much continuing his long-term tactic of arresting critics, holding them without charge or charging them with outlandish claims of "treason," releasing them, and then rearresting them—just as he had done to me.

Shultz even went so far as to declare the 1985 election "quite open" in his opinion and said the only problems he had heard about were some slight but not deeply troubling irregularities in the vote-counting process. These problems, he suggested, were most likely caused not by blatant and systematic fraud but by the fact that 75 percent of Liberians were illiterate and had not understood how to mark and cast their ballots. He called on the five or so remaining opposition leaders who still refused to take their seats in the National Assembly to end their boycott.

The howls of outrage from me and other Liberian exiles caused the Reagan administration to back down a little. Nonetheless, it was clear that those of us interested in bringing the full force of international condemnation against Doe were going to have a tough fight on our hands.

At the time I was quoted in newspapers as saying that Shultz had to be speaking either from serious misinformation or from ignorance. Clearly, though, he was speaking from neither but from a strategic concept that still placed Liberia in the United States' sphere of influence in those waning days of the Cold War. What's more, I believe Shultz was also articulating a common, if usually unarticulated, feeling about the limited possibility of democracy in Africa.

In July 1988, Doe survived yet another coup attempt, this one led by J. Nicholas Podier, another of the original band of soldiers who had led the April 12, 1980, coup. Podier later became Speaker of the National Assembly. Once again, Doe responded with a paroxysm of repression and violence, rounding up opposition leaders and imprisoning them, harassing and sometimes killing innocent civilians.

That same year the Reagan administration, exasperated not with Doe's human rights abuses but with his economic ones, sent seventeen financial experts to Liberia in an attempt to help Doe clean up the economic mess. This group of operational experts, known popularly as "opex," had but a limited impact on the pervasive and widespread corruption, lack of accountability, and dismal management. Shakedowns of ordinary citizens by soldiers and public officials were nearly an everyday occurrence for many Liberians. After only one year—despite its initial two-year commitment—the group packed up and returned home.

It was the beginning of the end of the U.S. commitment to the Doe regime. The world was changing: Reagan and Mikhail Gorbachev were meeting, reform was sweeping Eastern Europe, and the Berlin Wall would soon fall, signaling the end of the Cold War. Liberia's strategic importance to the United States was slipping fast. U.S. military aid to the country dwindled to zero by 1990, while economic aid was deeply slashed.

Right around this time a group of leading dissidents in exile, including myself, Amos Sawyer, Ezekiel Pajibo, Patrick Seyon, Levi Zangai, Tom Woewiyu, and others, formed an organization called the Association for Constitutional Democracy in Liberia (ACDL). Our goal was to advocate for change in Liberia by lobbying the United States and other governments to put more pressure on Doe, who was growing increasingly repressive. We petitioned the U.S. Congress and wrote letters to the editor. We worked with people interested in our cause. We demonstrated outside the Liberian Embassy in Washington. We did all this, and still Doe seemed only to strengthen his grip on power, to abuse more and more Liberian human rights, to further denigrate any hopes for democracy.

Then, sometime during the course of 1989, Tom Woewiyu came to the group and told us there was a military action being formed against Doe, headed by a man named Charles Taylor.

CHARLES TAYLOR was born in a suburb of Monrovia, the son of an Americo-Liberian father and a Gola mother. As a young man he journeyed to the United States to pursue an undergraduate degree at Bentley College in Waltham, Massachusetts. It was while in the United States that he joined the Union of Liberian Associations in the Americas and became active in that group's agitation against the Tolbert government.

In the wake of the Rice Riots, President Tolbert, hoping to appease the agitators and buy himself some time, invited a group of dissidents home to observe the government up close. Again at Tolbert's invitation, the group came to the Finance Ministry one warm day in January 1980 to get a firsthand view of the country's economic policies. I was finance minister at the time and thus naturally present at the meeting.

It was the first time I met the man who would go on to cause so much bloodshed and pain in Liberia.

That day Taylor struck me as a young and forceful man, charming and persuasive, with charisma to spare. Nonetheless, there were plenty of bright, charming, and ambitious young people in the country in those days; given the state of crisis in which we were operating, I had no time to pay any more attention to Taylor than to any of the rest.

The coup took place a few months later, and President Tolbert was killed. Taylor was still in the country at the time and was reported to be a close friend of Thomas Quiwonkpa, one of the leaders of the coup and Doe's right-hand man prior to their split. By proxy, he also gained the trust of Doe, who, after installing himself as head of state, appointed Taylor the head of the General Services Agency, the government's procurement arm.

Later, the relationship apparently soured. In 1983, the Doe people accused Taylor of embezzling nearly $1 million in government funds and Taylor fled the country, settling in the United States. Thomas Quiwonkpa had also fled into exile in the United States, and the two apparently reconnected there.

Soon, though, Taylor was arrested in Massachusetts on a federal warrant and imprisoned awaiting extradition to Liberia. He hired former U.S. Attorney General Ramsey Clark to represent him and to fight the extradition, which dragged out the process. During the year or so he spent in prison, according to newspaper reports, he was frequently visited by Quiwonkpa.

According to *The Boston Globe,* it was a Sunday night in September 1985 when Taylor asked a guard at the Plymouth House of Correction to bend the rules and let him pass from the north wing to the east wing so that he could play cards with a friend. The guard, for whatever reason, complied, then left his post for a few minutes. When he returned, Taylor and four other inmates were gone. They had used a hacksaw to cut an opening in the iron window bars and tied bedsheets together to climb down from the second floor.

Taylor disappeared. Two months later, Quiwonkpa led his coup attempt and I had the worst day of my life.

WHEN TOM WOEWIYU came to the ACDL sometime in 1989 and pushed for the group to support Taylor's armed resistance to the Doe regime, the question caused a rousing debate, with passionate opinions on both sides.

Some of us felt that militancy was an important ingredient in the struggle. We remembered how Malcolm X, by refusing to embrace nonviolence, made the white American power structure more willing to hear the words of the equally courageous and equally radical but nonviolent Dr. Martin Luther King, Jr. We also reasoned that apartheid would not have been abolished in South Africa without some level of force to facilitate that tremendous change.

Others in the group were less certain, feeling that they did not know Taylor well enough to trust his motives, though he certainly said all the right things.

For my part, I did not really know Taylor, having met him only that one time at the Finance Ministry and once again, when I was in Paris for work, through Tom Woewiyu. Tom had somehow tracked me down and said that he and Taylor needed to talk to me, so I invited them to have breakfast with me at my hotel. I was in the hotel restaurant, reading a book of speeches by Thomas Sankara, when they arrived. Sankara was, of course, the young, charismatic leader of Burkina Faso, who had implemented a revolutionary government committed to women's rights, improved education and health care, and the fighting of corruption. He was assassinated in 1987.

When Taylor and Woewiyu arrived, I put the book on the table. Taylor looked at it and said, "You mean people are still talking about that boy?"

"Of course," I said. "Everyone praises him for the things he did in his country."

Taylor said, with a dismissive gesture, "Well, he's gone now."

Years later, journalists and human rights experts would unearth evidence suggesting Taylor's deep involvement in Sankara's murder, but at the time this was not known.

"Let me order breakfast," I said. But Taylor said, "The money you spend to pay for breakfast you could just give to us."

I gave them what I had, and they left.

It was clear to me that whatever their plans, they were not going well at the moment if they needed the price of breakfast to keep on. But I was persuaded by Tom, whom I then held in high regard, to keep an open mind about Taylor. Tom, as the then president of the Union of Liberian Associations in America, had stood by me during my time in prison for the "idiot speech," having gone so far as to testify before the U.S. Congress on my behalf. Though others had deserted me in my time of need, he had not. I trusted him and felt I owed his passionate belief in Taylor at least the benefit of the doubt. I was also encouraged by a then friend and political ally, Grace Minor, a long-standing close confidante of Taylor.

On December 24, 1989, Taylor and a group of no more than two hundred exiled dissidents entered an area of northeastern Liberia more than 250 miles from Monrovia. The group engaged and killed an undetermined number of government soldiers and immigration officers and swiftly took control of the border town of Butuo.

Word of the incursion quickly reached Monrovia and Doe, who immediately dispatched two well-armed battalions to contain the rebellion. Under Doe's orders, the army moved brutally against villages in its path, indiscriminately killing, looting, raping, and terrorizing the population, which fled en masse to the safety of the forest or the sanctuary of neighboring villages in Guinea and Côte d'Ivoire. Most of the victims were of the Gio and Mano ethnic groups that inhabit Nimba County, a region regarded by the Doe government with hostility and suspicions since the abortive Quiwonkpa coup.

Doe also sent a message to opposition leaders when, on the night of January 4, 1990, his infamous death squad burst into the Monrovia home of my friend Robert Phillips and brutally murdered and dismembered him. Robert was a comember of the Liberian Action Party and the friend who had been with me on the day of the failed coup in 1985. He was a tough critic of Doe and had been one of those watching the ballot count at the Centennial Memorial Pavilion in 1985 who announced that Doe was losing—before Doe had his people take over the count.

Since the beginning of his reign, Doe had been able to crush every real and imagined coup attempt against his power. By overreacting so harshly this time, he may well have sown the seeds of his own eventual defeat—and for all the terror that came afterward.

In any event, word of Doe's scorched-earth campaign quickly reached us in America. When Taylor claimed that his men would protect the people of Nimba, some of us were inclined to give him the benefit of the doubt.

At the same time, I received a message from the U.S. State Department urging me and the members of ACDL to try to engage Taylor and his men, to encourage them to be disciplined and moderate in their tactics, to avoid civilian casualties.

At this point thousands of young villagers, especially among the Gio and Mano people, still angry about Doe's brutal reprisals in the wake of the Quiwonkpa coup and having watched their families murdered and their homes destroyed by government troops, began joining Taylor's forces. They joined not necessarily out of trust or approval of Taylor but in reaction to Doe's brutal rampage. This ragtag band of rebels, who called themselves the National Patriotic Front of Liberia (NPFL), drew increasing support from a nationwide population that neither knew nor necessarily trusted Charles Taylor, but that desperately hoped the end of Doe's ten-year reign of terror was finally at hand.

In the end, those of us who thought it necessary to try an armed resistance won the debate at the ACDL. The ACDL raised $10,000 to send to Taylor, who was then, we were told, somewhere in Côte d'Ivoire. The money was to be used to provide food for Taylor's troops and for the citizens of Nimba County, and we committed ourselves to doing more once we saw how the effort progressed.

It should be clear, however, that neither I nor any other ACDL member, other than Tom Woewiyu, was ever a part of the NPFL or had any knowledge of its plans.

A young man named Elmer Johnson, a former member of the U.S. Army and a very disciplined man, came to me and said he wanted to go to Liberia and join Taylor. He would help curb any tendencies toward abuse in the ragtag group of insurgents, would work to bring discipline and professionalism to the force. I encouraged him to go.

Later, I picked up my telephone one day and was surprised to find Taylor on the other end of the line. "We've got this tiger by the tail and we won't let go," he said. "I want you all to know that we're going to bring this thing down, and we need all the support we can get." I said okay and again urged him to be disciplined in his approach. The situation was volatile, and many, many people were in harm's way, including some of my own friends, such as Jackson Doe who, fearing for their lives under a rampaging Doe, had gone behind the Taylor lines.

Taylor said, "Yes, we know. We are arranging to take good care of them. Don't worry."

Indeed.

By May, Taylor's forces, the National Patriotic Front of Liberia, were claiming to be in control of much of the rural countryside, and the situation was escalating rapidly. Hundreds of thousands of refugees had already fled Nimba County and the surrounding countryside. I wanted to see for myself whom we were dealing with, so when an annual meeting of the African Development Bank was held in Abidjan in May 1990, I took the opportunity.

At the time Taylor was operating from a headquarters in the bush between Côte d'Ivoire and Liberia. I got some people to take a message to his camp that I wanted to come and see what it was all about, to see what he was doing and learn about his plans.

Journalists from the United States and Europe were swarming all over West Africa, trying to cover the emerging crisis in Liberia. A few British reporters were hanging around the ADB meeting headquarters, and when I said I was going to meet Taylor a few of them asked if they could come along. I said yes. So, after arranging security, we set off into the bush.

We drove to the border, and, after sending a message into the bush that we had come to see Taylor, we stopped to wait. We waited and waited and waited some more, for hours it seemed, until finally, toward late afternoon, a group of soldiers returned with a message from their leader.

"He says you can come," one of the soldiers said to me, "but you cannot bring any journalists or anyone else with you."

The reporters were not pleased with this development, of course, but none of us was about to challenge the decision. While they waited at the border in the car, I walked off into the bush, accompanied by Taylor's men. We had to cross a stream, and as we did I looked up and saw soldiers lined up on the other side, both male and female, all with blank, bloodshot eyes, just staring at us as we passed. After another half mile or so we reached the base: several small buildings surrounding a yard in which stood many huge and frightening guns, all watched over by hundreds of soldiers, soldiers as far as the eye could see. That Taylor was heavily guarded was very clear.

I was shown into one of the buildings and kept waiting for a while in a small anteroom. Eventually I was ushered into the main room, where Taylor sat, surrounded by bags of rice and deep in conversation with a Lebanese man. When I walked in, the conversation ceased. Taylor asked the man to excuse himself. Then, while I sat and listened, he delivered a long, complicated diatribe about all his plans.

He said he was tired of the Doe government's abuses and that he and his forces intended to redeem the country for all of the people. He said he was going to make calls to J. Rudolph Grimes, a former secretary of state, and Emma Shannon, a former associate justice, and would bring those good people back into the country to help rebuild what had been lost.

"And I'm going to deal with all these radicals like Amos Sawyer and Baccus Matthews, because they have destroyed the country with all of their radicalism," he said.

He talked some more along these lines, then said, "Come, let me show you how strong we are." We left the building and walked into the yard, where he proudly pointed out his stash of weapons. "I got all these guns from Doe," he said, "and I will use them against him."

When it was time for me to leave, he wanted to know what I thought of his plans and where I stood in relationship to them. I told him that at that stage I was not prepared to play a part, that I had come because I wanted to see for myself who he was and what he was about.

"Okay," he said and told his people to take me back to the meeting place, which they did. The journalists were still there, still hoping to interview Taylor, which, I think, they later did. I returned to Abidjan, very unsettled and musing hard on what I had seen and heard. The signals that Taylor, despite his grand rhetoric, was less interested in saving the people of Liberia and more interested in personal gain were apparent: the obvious dealmaking with the Lebanese man, the ill-gotten bags of rice.

I was also bothered not so much by what he had said as by what he had not. Taylor's talk was very unmeasured, it seemed to me, not at all grounded in the very real consequences of the path upon which he had embarked. There was no talk about how those big guns could just as surely be turned upon him and his men and the people of Liberia, and what that meant. There was no talk about the need to get the people on board with his movement or about how that might be done. There was no talk of that kind at all. There was only talk about what he would do once he had won, because Charles Taylor knew that he was going to win. And implicit in his confident, boasting rhetoric was the fact that he would do whatever it took to make his belief a reality. I left that meeting in the bush very unsettled about where we were going and what would happen along the way.

I returned to the United States, increasingly anxious about events in Liberia. By the time I reached America, the U.S. Embassy in Monrovia was advising American citizens to leave the country. In June, the U.S. State Department began evacuating some citizens, creating heart-wrenching scenes at the airport as Liberian parents put their American-born children on planes and sent them off to lonely safety in Abidjan or Washington or New York.

By July, two major developments had taken place that further served to doom whatever hope we had for a Taylor revolution.

First, Elmer Johnson, who had hoped to bring discipline, structure, and professionalism to the rebels, was abducted and killed by his own troops on orders from Taylor, reportedly out of jealousy because he was an Americo-Liberian gaining too much influence among the ranks.

Second, a commando named Prince Yormie Johnson had split from Taylor and become the leader of his own, smaller but equally danger-ous, rebel force. Johnson was to have led the western flank of the ad-vance into Monrovia and after his defection from Taylor announced his intention to combat both Doe and Taylor under the banner of the Independent National Patriotic Front of Liberia (INPFL). Suddenly Monrovia had not one but two rebel forces closing in.

Most of Doe's cabinet members, his chief of staff, the head of his presidential guard, and his top political adviser fled their leader; Doe holed up in the city with what remained of his loyal troops. His soldiers continued to hunt down and kill Gio and Mano citizens. They also at-tacked those of Americo-Liberian descent, whom they perceived to be hostile to Doe's largely Krahn regime. Missiles lobbed daily and indis-criminately from the Executive Mansion grounds destroyed whole sec-tions of the city, while death squads looted and burned with impunity. The headquarters of the United Nations was attacked, and dozens of people who had sought safety there were executed on a nearby beach.

The people of Monrovia were terrorized. Thousands fled from their homes to the safety of the borders, many of them eventually settling in Guinea, Sierra Leone, and Côte d'Ivoire. Those who remained strug-gled not only to dodge the fighting and the murdering armies on both the left and right but to sustain themselves, as all commercial activity ground to a halt. International aid agencies fled the country. Food con-voys were kept from traveling the roads or reaching Monrovia. Stores were soon looted; food stores disappeared. People began eating first the edible plants—wild cassava and cassava leaves, hibiscus, coconuts, palm cabbage—and then the inedible ones: swamp weeds and flower bulbs. They ate bush rats and frogs, snails and grubs.

One man told a reporter for *The Boston Globe*, "The dogs ate the dead, and we ate the dogs."

As the war grew worse and worse, some 500,000 Liberians, nearly 20 percent of the population, fled the country, becoming refugees and threatening to upset the entire West African region. As many as twice that number were internally displaced, creating a humanitarian crisis of staggering proportions.

The United States responded to these events by sending four warships to the waters off the coast of Monrovia, sending the hopes of many Liberians soaring to the skies. U.S. officials stated repeatedly that the ships and the marines on board were there solely to protect U.S. citizens. Testifying before the U.S. Congress on June 19, Assistant Secretary of State for African Affairs Herman J. Cohen made it plain: "The resolution of this civil war is a Liberian responsibility," he said. "There is no intention of using these forces for any political end; they are there to preserve American lives. I repeat, U.S. forces will not intervene to stop the fighting or to influence the outcome of this crisis in any way."

But so many people simply refused to believe it. Not only was Uncle Sam coming to save the day, he was there, in sight! Surely he'd stop the bloodshed, they cried. Surely Uncle Sam would not just sit back and watch his children die. The hope for rescue from America was epidemic in Liberia during those terrible months, but it turned out be bitterly false. The ships kept floating, the marines on board stayed on board, and Johnson and his forces and Taylor and his forces and Doe and his forces intensified their war.

It was around this time, around the second or third of July, that the BBC reporter Robin White called me at home to ask my thoughts about the events taking place in Liberia. White said something along the lines that Taylor was approaching the Executive Mansion and Doe was refusing to leave, which could mean that Taylor would end up bombing the mansion and even destroying it to bring Doe down. My response was "If they burn the mansion down, we will rebuild it."

Then the reporter asked me, "What if Taylor wins this war? What then?"

I said, "Well, we are nearing July fourth. On July fourth we'll drink champagne."

Of all the public statements I've made in my life—many of them brash, some of them harsh, all of them as honest and as forthright as I could make them upon uttering—those are the ones that have most come to haunt me and the ones I most regret. Over the years the word "Monrovia" came to be substituted for "mansion," so that I was made to seem to be advocating the wanton destruction of my own hometown, which was never my intention. Nonetheless, whatever the misinterpretation and whatever the situation in which I voiced my thoughts, it was a stupid statement to make. It was the kind of statement that has often led others to accuse me of being arrogant, and I have apologized to the Liberian people for making it. Still, I suppose I will never live it down.

At the time the situation was terrible: soldiers running rampant, people being beheaded in the streets. My thinking was: let the man win so that we can stop the bloodletting and restore some order. The same sentiment was being formulated by opposition leaders gathered in Freetown. Then, when the killing has stopped, we will deal with the victor. But, of course, at that time I did not know how truly bad the situation was, and just how bloodthirsty Charles Taylor would prove to be.

Soon we learned that Jackson Doe and several others who were supposed to be in Taylor's safekeeping had disappeared. No one knew what had happened to them, only that they had last been seen behind Taylor's lines, that Taylor had proclaimed them safe, that he was celebrating those great heroes of the nation. In truth, what was happening, as we soon discovered, was that Taylor was systematically killing anyone he considered even the remotest threat to his power. Jackson

Doe was killed. There are some who say that because we advocated for the protection of Jackson Doe we put him at undue risk, that he was killed by Taylor precisely because Taylor resented his popularity. But who can ever fully explain the motivations of a man such as Taylor?

Others were also killed, including the activist Moses Duopu and Gabriel Kpolleh. Likewise, thousands of innocent civilians were killed by Taylor's raw and undisciplined forces as they marched toward Monrovia. Homes and stores were looted, women raped, villages destroyed.

By the time we reached the end of July, I and the other members of the ACDL had withdrawn our support from Taylor. We realized, too late, that he was not part of the solution but part of the problem and, if anything, a worsening. So the ACDL went to work trying to mobilize the civilian component of Liberian leadership to see how we might be able to form an interim government that might rise above all warring factions: Taylor, Johnson, Doe.

In late July one of the single most horrific incidents of the war took place.

In the midst of the seesaw fighting that had gripped Monrovia for so long, St. Peter's Lutheran Church in the Sinkor section of Monrovia had been designated as a Red Cross shelter, a safe place of refuge and retreat. Some two thousand people had taken shelter there, pleading with foreign diplomats for help but receiving no concrete assistance.

According to reports by the reporter Mark Huband of the *Guardian,* on the evening of July 29, 1990, a group of about fifty government soldiers began driving around the church. A few hours after sunset, the soldiers burst in. For two hours they went through the church and the adjoining school, shooting everybody in sight. For three more hours they used machetes to hack the injured to death. In the end some six hundred people were slaughtered, including women with infants, old people and the disabled, children where they slept. Afterward the Red Cross flags the refugees had been given to hang on the fence as a sign to soldiers still fluttered in the breeze.

Speaking at an emotional demonstration in the following weeks, I tried as best I could to capture the enormity and horror of the situation and to assign blame where it lay.

"Let us make no mistake about it," I said.

66 THIS TOTAL disregard for human life so evidenced in this act is not an aberration. Liberian troops under the direct leadership of Sergeant Doe have demonstrated their cruelty time and time again. You and I know that those six hundred were not the first to die. We have closed our eyes and pretended not to see.

In 1980, Doe killed dozens, including the president. Some cheered. Some looked the other way. Most ignored and went on about their daily lives. In 1981, Doe killed again, this time six of his own colleagues. The other colleagues and the rest of us shrugged, pretending this savagery was in the interest of the nation. Between 1981 and 1984, several others accused of plotting to overthrow the government were killed. We did not blink, believing it to be within their own family. As we say, John's palm oil on John's rice.

Now we face the truly astounding situation in which an international sanctuary at a church has been attacked and hundreds, mainly women and children, killed at the hands of those who can be called no more than beasts. We must therefore demand a response from all those who by their continued combat prevent relief getting to Monrovia. They must stop the combat now. They must put national interest above personal ambition. And we must all look at ourselves. 99

Those were my words, but even as I spoke them I was beginning to fear, as were many of us, that there were going to be no saviors in this situation, no good guys to whom to turn.

By August, Prince Johnson and his troops had taken control of Bushrod Island, just over the bridge from Monrovia. Johnson announced that all foreigners in the capital would be arrested, a threat he said was intended to provoke international intervention in the war.

Like many Liberians, Johnson said he would prefer it if the United States stepped in, but he would settle for the West African community. In a way, his tactic worked: a day later Sea Knight helicopters landed at the U.S. Embassy compound and at two nearby communication sites and marines began evacuating American personnel. Johnson, still trying, then seized a handful of foreign citizens, including one American, as hostages. The United States still refused to engage any troops on the ground. A few days later Johnson released the hostages.

But though neither the United States nor the United Nations responded to the cries of Liberia, West Africa did.

The Economic Community of West African States (ECOWAS) was formed in 1975 in Lagos, Nigeria. Fifteen nations—Benin, Côte d'Ivoire, Gambia, Ghana, Guinea, Guinea-Bissau, Liberia, Mali, Mauritania, Niger, Nigeria, Senegal, Sierra Leone, Togo, and Burkina Faso (joining as Upper Volta)—signed a treaty intended to promote economic self-sufficiency and overall development of the West African region. Cape Verde joined the union in 1977. Mauritania left in 2002. Although extremely slow progress has been made in the effort to create a single large trading bloc in West Africa, the member states remain hopeful that a single monetary zone with a common currency will eventually be formed in the region. Likewise, efforts are under way to enhance regional cooperation through connectivity in electricity, gas, and roads.

It was this collection of West African governments that acted in an unprecedented manner to meet the disaster unfolding in Liberia. During the months of July and August, at meetings held among themselves and with the warring factions, under the auspices of then chair Sir Dawda Jawara of Gambia, ECOWAS organized a Mediating Committee that instituted a peace plan for ending the conflict. The plan called for an immediate cease-fire, an all-party conference to select an interim government, and free and fair elections to be held within twelve months.

Most critically, the plan also called for the formation of a regional military force to enforce the cease-fire, a force to be known as the ECOWAS Monitoring Group, or ECOMOG. Thus a five-nation,

4,000-member force was assembled in Freetown, led by Nigerian armed forces and under the command of Lieutenant General Arnold Quainoo, a Ghanaian.

Not all of the ECOWAS countries were happy with the decision of the Mediating Committee to send troops to Liberia. Many of them cited charter restrictions of the Organization of African Unity against interference in the internal affairs of sister states. Differences among the ECOWAS states seemed to fall along linguistic lines, with the Francophone states supporting the noninterference stance promoted by Burkina Faso and Côte d'Ivoire and the Anglophone states lining up behind Nigeria and Ghana to support the intervention and the interim government.

The peace plan also called, in effect, for Samuel Doe to relinquish the presidency to the head of the interim government, a person who would not be any of the leaders of the warring parties. Not surprisingly, both Doe, who was still stubbornly clinging to power, and Charles Taylor, who had proclaimed himself president of Liberia, rejected this idea. Only Prince Johnson, who controlled most of Monrovia, said he would welcome the force.

Taylor claimed that Nigeria and Ghana, which were supplying the bulk of the forces, were both sympathetic to Doe and had supplied him with arms. He threatened to attack any international soldier who set foot in Liberia. Whatever the truth of his assertions, it was also clear that Taylor wanted victory and glory for himself, not for any international force.

The last time I talked to Tom Woewiyu, who was now Taylor's minister of defense, I said to him, "We are going to Banjul. It would be good if the NPFL joined the talks so that we can form a government of national unity and maybe that way we can all force Doe to step down and save the nation."

Tom said, "They are thinking about an intervention force, and if they do intervene for us, that will be war. If that happens, we are not part of any peace."

On August 24, 1990, I was boarding the plane at Dulles International

Airport and listening to the BBC when I heard the news: ECOMOG had landed in Monrovia. The forces had been met at the port by Prince Johnson, who vowed to support them, but the port and the airport were shelled by Taylor's men.

I immediately called my friend Clave.

"I'm on the way to Banjul," I said to her. "But I'm afraid that we are not going to succeed." It was very clear from Taylor and his people that if ECOMOG landed, it was war.

I went on to Banjul, fearful that the ECOMOG landing would undermine the success of the conference and the peace process itself. Taylor had claimed he would accept a small, civilian monitoring group to oversee the cease-fire but not armed soldiers. His NPFL refused to send delegates to the meeting in Banjul.

Nonetheless, the conference elected Amos Sawyer as interim president. I myself did not seek the office because I was not prepared and because I did not think the group assembled there would elect me. I knew Amos Sawyer would do a fine job and that the important thing was to get a nonmilitary person into office if at all possible. As head of the major party, the Liberian Action Party, my signature was the first one on the document that elected Sawyer. The Lutheran bishop Ronald Diggs was chosen vice president.

It was initially decided that Winston Tubman would be minister of foreign affairs so that he could go out and use his international connections to rally people to this new government. One of the people close to Sawyer said to me that he wanted me to be in the government and that I would be able to hold a portfolio. He conveyed this desire to me through Byron Tarr. This bothered me, the hint that things were returning to the old way of doing things, of personal advancement over political life. I met with Sawyer and said, "If you are going to start to repeat the same kind of government we had before, I have no interest in it."

The next morning, I learned that they had met during the night and decided that, in fact, Tubman would not be minister of foreign affairs. Instead they chose Baccus Matthews, in what was clearly intended as a signal that they were going the radical route, that the progressives of the 1970s were taking over the government. I decided it was best for all concerned if I stepped back from things at this point, so I told Sawyer that I had a job, that I had not come to Banjul to look for a job, and that, while I wished them all well, I was not prepared to be part of the government. I remained at the conference for the rest of the day, then packed my bags and left. I flew back to Washington and waited to see what events would bring.

# CHAPTER 11

# ECOMOG

DESPITE THE events in Banjul, the fighting in Monrovia continued for several more months, albeit with some restraint due to the presence of ECOMOG. With Johnson's forces controlling much of the capital city and Taylor's forces commanding the surrounding countryside, Samuel Doe became a virtual prisoner in the well-fortified Executive Mansion. Still, he continued to refuse the United States' call to resign in the interest of peace.

Then, reportedly, he changed his mind.

On September 9, 1990, the ten-year reign of Doe, which had begun in a fit of violence, ended in the same brutal way. Prince Johnson and his troops captured Doe after a shootout with Doe's men at the ECOMOG headquarters. Doe had reportedly come to the headquarters for a meeting intended to secure his safe passage out of Liberia. Instead he was wounded in the legs, captured, and taken to Johnson's headquarters, where he was brutally tortured—beaten, his ears sawed off—and eventually killed. A videotape of his torture, mutilation, and death was quickly and widely circulated throughout the country and even the greater international Liberian community. As much as we fought Doe over the years, as much pain as he caused, he did not deserve this kind of end. No person does.

Sadly, Doe's death did little to stop the fighting. If anything, it intensified the battle. Convinced of their imminent slaughter, his troops moved viciously and violently against the population. In the rampage hundreds of innocent civilians were killed; reporters from UPI and Reuters would later report finding hundreds of skulls and bones at the end of a runway at Spriggs-Payne Airfield. Office buildings and homes were torched, utility poles uprooted, vehicles overturned and destroyed. Whole sections of Monrovia went up in flames, including the offices of the *Daily Observer,* a leading newspaper often critical of Doe. It was days before ECOMOG was able to bring Doe's troops under control.

At the end of November 1990, ECOWAS held another summit on the crisis, this time in Bamako, Mali, and with the participation of the OAU chairman. Hopes rose when Taylor actually consented to attend, and a comprehensive cease-fire agreement was reached—and signed—by Taylor's NPFL, Johnson's INPFL, and what remained of Doe's Armed Forces of Liberia (AFL). Unfortunately, the summit was widely believed to have weakened the interim government of Dr. Sawyer, because it failed to firmly recognize that body. For his part, Taylor categorically denied the agreement gave the interim government any legitimacy at all.

A week or so later, the guns fell silent in Monrovia, and the people of Liberia began slowly and tentatively trying to survive. The city was in ruins, with many people, especially children, severely malnourished or critically ill. There was much work to be done burying the dead; reporters who reached St. Peter's Church found the bodies of the massacred still piled in the pews and in the yard.

Sawyer set up government in the once-grand Ducor Hotel, which overlooks Monrovia from a hilltop, surrounded by squatters and refugees who had come into the city during the war. He began trying to govern, a nearly impossible task.

There was little progress toward a political settlement of the crisis. Taylor, who had set up headquarters in Gbarnga and whose forces then controlled more than 90 percent of the country, wanted to be president and was unprepared to gamble on the likelihood that the Liberian people would agree with his desire via the ballot box. Even had Taylor been willing to relinquish his claim to the presidency, his supporters

weren't likely to let him. They believed they had made personal sacrifices and were thus entitled to the benefits of the conflict. Taylor's
commandos, battle-fit from three years of hard training in Libya and
Burkina Faso, were adamant in voicing their right to claim the spoils of
victory. They proclaimed that the time of the Nimba people had come.

Common wisdom had it that Taylor agreed to negotiations only because ECOMOG and Johnson had kept him from seizing Monrovia.
What followed was a year-long series of ECOWAS-brokered attempts
at peace agreements that largely failed.

First came the Lomé Agreement in February 1991: representatives
of the three warring factions attended and signed agreements for the
encampment of troops under ECOMOG supervision. Togolese President Gnassingbé Eyadéma called Taylor and Johnson to a meeting to
work out arrangements for a proposed National Conference.

That conference, the All-Liberia National Peace Conference, was
held one month later in Virginia, Liberia. More than 150 delegates
from major political parties, the warring factions, various interest
groups, and assorted county representatives attended. Taylor, breaking
a previous promise, did not. He did, however, send representatives.

Nonetheless, the conference was held. It endorsed Sawyer as president of the interim government, created two vice-presidencies for
Johnson and Taylor's groups, and inaugurated a 51-member, unicam-
eral Interim National Assembly. The NPFL representatives pulled
out of the conference before it ended, complaining about the group's
unwillingness to entertain Taylor's proposal for a three-person ruling
council—consisting of himself, Sawyer, and a neutral third party—
which would run the country until elections were held.

In July and again in September and October, conferences were
held in Yamoussoukro, Côte d'Ivoire, under the guidance of President Houphouët-Boigny. The heads of state of Burkina Faso, Gambia,
Guinea-Bissau, Mali, Senegal, Togo, Nigeria, and Ghana attended, as
did Taylor and Sawyer. The result, known as the Yamoussoukro Accords, mandated disarmament of the various warring factions.

But in the midst of all this activity, NPFL forces were clashing with Sierra Leonean troops along the Liberian–Sierra Leonean border, threatening that country's stability. At the same time, refugees seeking safety flooded back into Monrovia, swelling the city's population to 800,000 by May—nearly one third of the prewar population of all of Liberia.

Throughout this time the people of Liberia were trying to pick up the pieces of their shattered lives. In Monrovia aid agencies that had fled during the vicious fighting returned to find a city devastated by war, with severely malnourished people, diseases such as malaria and other sicknesses prevalent, and bodies still rotting on the beaches and in killing zones such as St. Peter's Lutheran church.

Monrovia, controlled by ECOMOG, was largely stable and safe. A song in praise of the West African peacekeepers sprung up and quickly gained popularity: "Thank God for ECOMOG!"

All of this I watched from America, wondering what the outcome would be. There is no doubt that ECOMOG initially saved lives by stepping into the crisis when other international groups—the United States, the United Nations—declined to intervene. For that, all Liberians are grateful. It is also without doubt that the Liberian war cost the ECOWAS countries the precious lives of thousands of their peacekeepers and millions of dollars in funds.

But the ECOMOG intervention was also a flawed operation in some ways. Much critical analysis has been written, and there remains considerable dissent about not only the legality and methods of the intervention but about whether ECOWAS shortened or actually prolonged the conflict in the long run.

Writing in the *African Studies Quarterly,* the analyst Christopher Tuck cites, among other problems: lack of proper funding, equipment, and troops; suspicions and conflicting interests among the ECOWAS member countries; and lack of clarity in the mandates given to the force. This lack of clarity—whether ECOMOG should restrain itself to peacekeeping or focus on peacemaking—as well as Taylor's vicious tenacity, led the group to ally itself with one or another of the warring factions, which greatly undermined its credibility. Also contributing to the problem were the many incidents of corruption and theft by individual members of the (often underpaid) ECOMOG forces, which, as Tuck wrote, led to the "local joke that ECOMOG was an acronym for 'Every Car or Moving Object Gone.'"

Most critically, I believe ECOMOG could have ended the fighting much earlier than it did. Yet, whether through corruption, ineptitude, or confusion, the force was never able to follow through on its initial victories against Taylor's forces. More than once we received reports of ECOMOG forces having Taylor on the ropes, bombing his base, and advancing strongly. I once received a telephone call in the United States from Grace Minor, one of Taylor's top associates, begging me to "make ECOMOG stop"—as if I could or would. And yet it seemed as if somehow Taylor would always find a means of escape, whether by bribing some of the on-the-ground ECOMOG commanders or by other means.

My feeling, at that point and now, was that once ECOWAS intervened, it should have brought an end to it all as quickly as possible. Instead the group seemed caught in endless negotiations with all parties, never quite certain which group it really wanted to back. One minute there would be a strong ECOMOG commander and one would hear they were bombing Gbarnga, where Taylor was headquartered; the next minute you'd hear they had let up on the bombing—in effect, if not in intention, giving Taylor time to regroup.

It is important, when assessing the role of ECOMOG in the Liberian crisis, to keep in mind the fact that regional peacekeeping in national conflicts was, at the time, a relatively new phenomenon. ECOWAS's leading role in the conflict was based on the premise that countries in a region could achieve certain objectives more efficiently through collective rather than individual efforts. It made sense that one major objective of any such regional partnering should be the maintenance of regional stability, since it was clear, from the moment Taylor crossed into Liberia to launch his war, that the conflict contained the potential to spill over into neighboring states. Which later it did.

It is also important to remember that ECOWAS took on the

roles—first of peacekeeper, then of peacemaker and chief negotiating unit—with an organization that possessed neither the structure nor the mandate sufficient to performing the tasks at hand. Unlike the United Nations, which commands both worldwide resources and broad, if not universal, recognition as a neutral agency, ECOWAS had not had time to establish either credential.

This was a critical point. When ECOMOG first intervened in the conflict, Charles Taylor labeled the force a thinly veiled attempt by Nigerian President Ibrahim Banbagida to rescue his friend and colleague President Doe. In response to the ECOMOG landing, Taylor's NPFL targeted Nigerians, Ghanaians, and other nationals of the ECOMOG countries, detaining or even executing hundreds of people caught in NPFL territory.

Taylor's accusation not only tarnished ECOMOG's neutral role but had a direct impact on the refusal of the warring factions to disarm and reach a political settlement. Taylor constantly maintained that ECOMOG could not be trusted because the force was arming his enemies. Unfortunately, credence was given this claim in 1992, when Taylor's brutal offensive against Monrovia compelled ECOMOG to seek the assistance of other warring factions in defending the capital. Sources in the U.S. State Department later confirmed that ECOMOG had been involved in either directly supplying or facilitating the transfer of arms to several warring factions. Some observers believe this conduct only served to prolong the crisis.

On this point, at least, the United Nations must share some blame. Suspicions regarding the neutrality of ECOMOG could have been reduced had the United Nations, early in the process, implemented Chapter 8 of its own charter, which gives the Security Council the authority to, in effect, deputize regional actions. Clear and consistent U.N. approval of ECOMOG would have conferred a certain legitimacy upon the monitoring group that it lacked.

Still, it is fair to say that ECOMOG jeopardized its neutral stance by failing to take action when ECOMOG troops took part openly in the looting of Liberia. Cars, trucks, generators, computers, metallic railings, scrap metal, and a host of other goods were stolen and transported out of the country on the same ships that brought in ECOMOG troops. Nor did Nigeria react effectively when consistent reports emerged that the Nigerian ECOMOG contingent was actively supporting one of the factions, the Liberian Peace Council.

There were other problems with ECOMOG, both in its creation and its actions on the ground. The noted Liberian scholar D. Elwood Dunn has documented some of these in studies dealing with the role of ECOWAS and ECOMOG in Liberia.

Before ECOMOG even reached Liberian ground, ECOWAS should have done much more to develop a consensus on the need for a military intervention, both among Liberians and among the regional partners in general. This consensus building was critical, especially in light of the fact that intervention was contrary to the mandate and charter of both ECOWAS and the OAU.

As it was, there was a general failure to involve key regional states—including Burkina Faso and Côte d'Ivoire, both known to be supporters of the Taylor-led NPFL—in the decision to intervene. Of the sixteen member states of ECOWAS, only seven participated in the decision to send an intervention force. The failure to actively involve Côte d'Ivoire in the decision helped to exacerbate tensions between Francophone and Anglophone states in the region.

Further unnecessary political tension was created in the country when ECOMOG landed in Liberia in August 1990, before a cease-fire was reached. More tension was generated when officials moved ahead to establish the Interim Government, before meaningful consensus-building work was done. The selection of Amos Sawyer, rather than an individual perceived to be closer to the center of Liberian politics, also served as a bone of contention.

Most of all, what was missing once ECOMOG had landed and secured Monrovia was a clear and unequivocal consensus about what needed to be done. What was missing was a forceful statement, both in word and in deed from ECOWAS. That statement should have been: "We're in there to clean this thing up, we're going to remove

Taylor from the scene, we're going to remove Johnson from the scene and there's an interim government that we're going to back and everybody else is simply out."

Instead what took place was a series of endless meetings. Meanwhile, Taylor, who took seriously none of the peace conferences, set up shop in Gbarnga and ruled with great pomp over what came to be called Greater Liberia, or "Taylorland."

He installed himself as president and appointed a cabinet. He set up a radio station and had American soul music broadcast throughout the day. He had his Boston valet dress him in finery. He printed up stationery with a presidential seal. He collected taxes from all over, including Firestone. He created his own currency, called Liberty, and demanded that anyone doing business inside the territory he held use it or else.

Taylor also used the cease-fire time to begin exporting some of the valuable national resources in the land under his control. Iron ore, hardwoods such as mahogany and ebony, and, most of all, diamonds were stripped from the countryside and sold on the international market. The money he received for these goods he used to buy more weapons for his troops.

And so for two confusing, brief, and only relatively peaceful years, Liberia had what amounted to two governments on the ground.

# CHAPTER 12

# UNDP and Rwanda

In mid-1992, while still at Equator Bank, I was recruited by the United Nations to become assistant administrator and director of the Regional Bureau for Africa of the United Nations Development Programme with the rank of assistant secretary-general of the United Nations.

The U.N. system was trying very hard at the time to attract more women into roles of leadership. When the change in leadership at the Africa bureau took place, UNDP made a special effort to look around for a qualified woman who possessed both the political and professional skills to do the job. William Draper III, who was then the administrator of UNDP, had a large part in recruiting me. During the interview process I broke tradition by refusing to go to New York and lobby the various African ambassadors in residence. I said—and it even got into the papers that this was my position—that if I got the job it would be on the strength of my qualifications, not on the effectiveness of my lobbying skills. I got the job, becoming the first woman to hold the post.

The new position meant I could no longer be involved in Liberian politics. I had to disengage myself, and so I did. Though I kept in touch with events in my homeland, both through news sources and by talking to friends, I refrained from any active participation in political issues. It would be nearly five years before I dipped my toes back into those waters.

The UNDP, headquartered in New York City, is a significant multilateral source of development assistance in the world. The agency provides training, expert support, and funding to developing countries as they seek to develop local capacity, defined by the agency as the ability of individuals, institutions, and societies to perform functions, solve problems, and set and achieve objectives in a sustainable manner. The UNDP is a crucially important organization for Africa, and I wanted to use my term as head of the Africa bureau to not only raise awareness in the greater United Nations of African needs but to raise the UNDP's profile in Africa itself.

In terms of work environment, UNDP was a different world from the private sector. At UNDP, and the United Nations itself, even though professionalism is very much part of the job, diplomacy and politics prevail. In the UNDP environment, relationships—with governments and government officials, with the donor communities that provide the funding, and with the various institutions resident in New York—are far more important than anything else.

I was able, I believe, to bring a bit more professionalism to the bureau during my tenure. I was not seen as an insider, because I took positions that reflected my own experience working not only in government but also in the private sector. These were not always popular, but I was respected for them and for my performance. My work habits were the same as they always had been, and I became known for leaving the resident representatives messages at eleven or twelve at night. The next day people would ask, "Who else got a love note at midnight?"

I was also able to recruit and put into positions of authority a good number of women, some as resident representatives and some as deputies, which prepared them for eventual leadership. Moreover, the general promotion of women was a big issue in the United Nations at the time, and I traveled all over Africa and in particular to South Africa and Botswana to promote events and programs that helped women's development.

Boutros Boutros-Ghali was secretary-general of the United Nations at the time of my recruitment. A few years later, Kofi Annan took over that very important position, and, because of the connection between us as sub-Saharan Africans, I was drawn closer to the seat of power at the United Nations, increasingly called to the secretariat for meetings and appointed to various advisory committees. Kofi Annan became someone I could talk to, and he even came to dinner at my small apartment in New York. You can believe that did not go unnoticed!

Working at UNDP gave me a much broader perspective on Africa than I'd had before. I met with many of Africa's leaders at the time and cultivated strong relationships with quite a large number of them. Some of these African presidents, including Ugandan President Yoweri Kaguta Museveni, would come to play a pivotal role in advising how I financed my first presidential contest.

Perhaps the person I liked most among them, however, was Julius Nyerere. Nyerere, first president of the independent Republic of Tanganyika, oversaw not only the peaceful transition of Tanganyika from British colonial rule to independence but also the union of the country with the islands of Zanzibar and the creation of the Republic of Tanzania. Despite all of this, he was a man who lived his life with great moderation and simplicity and even greater integrity. His commitment to his people and his country, to Africa and pan-Africanism, has served as an inspiration to me. Some could call his record mixed because of the failure of his socialist economic programs, especially the collectivist agricultural system known as Ujamaa. But Ujamaa means "extended family" in Swahili, and whatever his shortcomings, Nyerere left his people unified. That is no mean feat in Africa—or anywhere in this increasingly fragmented world, for that matter—and is not to be easily dismissed. I also like the way he called for development of people in the rural area, which was and remains the heart of the continent. Nyerere was a good leader in my book, and it was an honor and a pleasure to get to know him.

Then, of course, there is Nelson Mandela. I first met Mr. Mandela at an OAU summit; I asked specifically to meet with him. He is, after all, the icon for all of us in Africa, and no one who considers herself a good revolutionary, or a good politician, in Africa would ever be happy unless she had met Nelson Mandela and had a chance to exchange

views with him—and have a photo opportunity. He and I spoke together for ten minutes or so. For my part, I told him how much all of us
Africans respected and admired him personally as well as the example
he had set for those of us who aspired to political leadership. In return
he talked a bit about the role Liberia played in support of the liberation
movements in South Africa and elsewhere.

At UNDP I organized and chaired roundtable conferences on development in many African countries, including one on Angola in
1995. That conference, held in Brussels, focused on gathering aid to
help Angola recover from twenty years of civil war. We were able to
bring together all major elements of the Angolan community, including, most critically, President José Eduardo dos Santos and his rival
Jonas Savimbi, the leader of the National Union for the Total Independence of Angola (UNITA), who had signed a peace accord.

Delegates to the conference pledged $1 billion in aid for Angolan
reconstruction, and the meeting was praised in the United Nations and
by the White House. Unfortunately, when the power brokers in question returned to Angola, the agreement fell apart. Savimbi, who had
been offered the vice presidency, turned it down, and war resumed. Savimbi was killed in 2002 during a conflict with government troops.

During our brief meeting I found Savimbi to be a stoic, somewhat impenetrable man. He spoke little but sat still and quiet while
his people worked the negotiation process. Although he had people
around him who seemed to be trying hard to do the right thing, his
own demeanor suggested a person capable of great ruthlessness. I had
a slight fear of him.

I was head of the Africa bureau in 1994, when the genocide unfolded in Rwanda. As we now know, that horrific event began the
evening of April 6, when a plane carrying Rwandan President Juvé-
nal Habyarimana and Burundian President Cyprien Ntaryamira, both
Hutu, was shot down as it approached the airport in Kigali. Hutu ex-
tremists, bitterly opposed to the Arusha Peace Accords and suspecting
that Habyarimana was finally about to implement it, are believed to
have been behind the attack.

Within hours of the crash, Rwandan armed forces and the Interahamwe militia set up roadblocks and began the systematic slaughter of Tutsi and moderate Hutu. The United Nations Assistance Mission for Rwanda (UNAMIR), created in October 1993 to keep the peace and assist the governmental transition, tried at first to intervene and stop the killings. The commander in charge, Lieutenant General Roméo Dallaire, a Canadian, pleaded for more troops and a mandate to stop the bloodshed, but instead the U.N. Security Council, urged on by the United States, voted to cut his troops from 2,500 to 450 ill-equipped men.

For the next hundred days the world stood by while eight hundred thousand people were viciously and systematically slaughtered. As the development arm of the United Nations, our agency had no role in the terrible events taking place; that role belonged to the Security Council and the Secretariat's office. At UNDP we stood with the rest of the world and watched those terrible events unfold.

Only when the rebel Tutsi force the Rwandan Patriotic Front took control of the country and installed a new government did the bloodshed cease. (Or rather, the genocide ceased; the bloodshed continued, sporadically, in the refugee camps and as the Hutu militia continued to lead incursions against the new government.)

That August, I was among the first group of outsiders to visit the scene of the mass murders as part of a U.N. mission to assess the situation. We flew by helicopter, visiting most of the areas worst hit by the genocide, including the capital of Kigali, Byumba in the north, Ruhengeri in the northwest, and Gikongoro in the south. It was a sight I will never forget and that I hope to God never to see again. Decomposing bodies, bloated from the heat and buzzing with flies, lay strewn across the countryside. In some places the sight of bodies rotting in the sun was so horrific your inclination was to turn your head, only there was nowhere to turn without seeing another person, a former person, dead and mutilated: flung along the road, dumped into a ditch, floating in a stream. The machete was the killing machine of choice for many Interahamwe, and that was obvious from what we saw. The Hutu commanders urged their followers to "send the Tutsi back" to Ethiopia (from which it is believed the group originally immigrated) by the rivers, and that was precisely what many tried to do: we saw rivers and streams so choked with bodies their levels rose. It was death and destruction on an almost unimaginable level. And everywhere there was the stench.

UNDP immediately mounted an effort to help the country rebuild. The following year, I chaired a roundtable conference on development in Rwanda during which we mobilized some $700 million in reconstruction funds for the country. The plan and the funds were aimed at rebuilding the country's economy, infrastructure, administration, and legal system.

Unlike Angola, Rwanda took the money and the energy mobilized during that conference and did the right things to rebuild its society. Paul Kagame, now president, was defense minister at the time and very involved in what we were trying to do, which was to present Rwanda to the international community and convince it that this was a country with promise, a country that, even with all the cleavage in its society and the terrible lingering effects of the genocide, had a chance to mend. As a result of that effort, which was successful, President Kagame remains a good friend today. Since becoming president myself, I have had the honor of visiting Rwanda and being a part of its national observation in remembrance of the genocide.

Why was Rwanda able to leave the conference table and move forward, while Angola was not? Why are some countries able, despite their very real and serious problems, to press ahead along the road to reconciliation, recovery, and redevelopment while others cannot? These are critical questions for Africa, and their answers are complex and not always clear. Leadership is crucial, of course. Kagame was a strong leader—decisive, focused, disciplined, and honest—and he remains so today. I believe that sometimes people's characters are molded by their environment. Angola, like Liberia, like Sierra Leone, is resource-rich, a natural blessing that sometimes has the sad effect of diminishing the human drive for self-sufficiency, the ability and determination to maximize that which one has. Kagame had nothing. He grew up in a refugee camp, equipped with only his own strength of will and determination to create a better life for himself and his countrymen.

Witnessing the aftermath of the genocide in Rwanda had a profound effect on me. This was, after all, Africa, and these, in a way that felt increasingly real to me as I worked throughout the continent, were my brothers and sisters. To see them succumb to such overwhelming brutality and violence was a terrible thing.

I have often wondered, too, about the anguish that Dallaire must have suffered. In a newspaper report by Rosemary Goring of *The Herald* (Glasgow), Dallaire tells about the day he broke in the midst of the genocide. He was en route to a meeting with the factions in an effort to stop the bloodshed, when he had to get out of his vehicle to cross a makeshift bridge on foot.

**❝** UPSTREAM, HE could see soldiers fishing bodies out of the water. He was used to such sights and did not flinch. But as he stepped onto the bridge, he noticed clothes caught between its slats. He looked into the water and his gaze was met by the empty-eyed stare of countless bloated, half-naked corpses, on which the floating bridge was resting. He was walking on a mattress of rotting flesh.

At that moment, the protective emotional screen Dallaire had built around himself shattered. 'My stomach heaved and I struggled for composure. I couldn't bear the movement of the bridge, up and down on the slaughtered hundreds.' **❞**

The issues that led to the violence in Rwanda and Burundi were far from new. Since the earliest days of the region, there had been conflict between the majority Hutu and the minority Tutsi. However, most evidence suggests that the two groups had lived a largely peaceful, if somewhat hierarchical, coexistence—speaking the same language, sharing customs and living areas, intermarrying—until colonialism arrived in 1897.

First the Germans and later the Belgians viewed the Tutsi and Hutu as two distinct peoples—and declared the Tutsi the superior group. The Tutsi embraced the designation, for decades enjoying better jobs, better education, and better economic opportunities than their Hutu countrymen. Time and time again humanity has failed to appreciate the deep anger and resentment, the very real and very dangerous pain that comes from being told you are inferior. In Rwanda, that resentment festered even after the country declared independence in the factions and the Hutu took power. The Tutsi who remained in the country became national scapegoats, blamed for every problem and every ill.

The situation festered and festered some more. Successive periods of ethnic killings took place over the decades, yet nothing was done to address the huge cleavage between the populations that underlay the violence.

Anyone paying attention could not have failed to see some chilling similarities between Rwanda and Liberia. Our situation was not of equal proportion, of course. For one thing, Liberia did not have the same huge pressures as Rwanda regarding land. Rwanda is the most densely populated country in Africa. Many Hutu said they feared the Tutsi-led RPF would take their individual plots of land. In some ways, the brutality in Rwanda was as much about very poor people manipulated into what they saw as a defense of their sole source of support as it was about ethnic hatred.

Nonetheless, I saw similarities to Liberia, and they worried me. The message was the same: when one allows a cleavage to go for so long without addressing it forcefully and frontally, one sets the stage for disaster.

In 1998, the OAU set up a high-level panel to investigate the genocide. I was one of two women appointed to the seven-person committee, the other being Lisbet Palme, chairperson of the Swedish Committee for UNICEF.

Two years later we issued our report *Rwanda: The Preventable Genocide.* In it we took a deep look at the roots of the conflict, and took sharp aim at the international community, including the Anglican and Roman Catholic churches, for failure to either prevent or halt the events.

At a news conference introducing the report, one of our members, Stephen Lewis, a Canadian member of the panel, said the French behavior was particularly indefensible.

"We repudiate the position of the government of France, the position that asserts they had no responsibility," he said. "They were closer in every way to the Habyarimana regime than any other government. They could have stopped the genocide before it began. They knew exactly what was happening."

Our investigations also made clear that the French peacekeeping mission eventually sent to Rwanda, ostensibly to set up a safe zone for civilian refugees, in fact created a kind of clear exit way for the Interahamwe killers and Hutu government officials who had masterminded the genocide. They had fled in safety to neighboring Zaire, now Congo, set up shop in the refugee camps there, and begun using them as bases to launch attacks on the new Rwandan government. The resulting Great Lakes refugee crisis continued for years, threatening the entire Central African region.

Our report called for France, Belgium, and the United States to pay a significant level of reparations to help rebuild the devastated country and for Rwanda's international debt to be canceled. We wanted to make sure the genocide was not forgotten or consigned to history because it had taken place in Africa.

I was still at the United Nations in December 1996 when Kofi Annan became secretary-general, the first person from a black African nation to serve in that prestigious role. As an African, I was especially elated at his promotion to this most important job. Annan was, and remains, a seasoned and impressive diplomat who also possessed great compassion for the world's struggling peoples and offered great resistance against those who would trample human rights. During his term he issued a "Call to Action" to address the HIV/AIDS pandemic in developing nations, supported sending peacekeepers to Darfur, and addressed international terrorism.

In accepting the 2001 Nobel Peace Prize on behalf of himself and the United Nations, Annan said, "What begins with the failure to uphold the dignity of one life, all too often ends with a calamity for entire nations." Annan warned that genocide begins with the killing of one person—not for what he has done but because of who he is.

Unfortunately, Annan's legacy will be forever tarnished by the genocide in Rwanda. Annan was U.N. assistant secretary-general for peacekeeping operations at the time the genocide unfolded, and he himself has admitted that he could and should have done more both to prevent the genocide and to stop it once it had begun.

In particular, in January 1994 Roméo Dallaire sent a cable to General Maurice Baril, military adviser to then–UN Secretary-General Boutros Boutros-Ghali, warning about a cache of hidden Interahamwe weapons, which he feared were being stockpiled to slaughter Tutsi should a conflict begin. It was Annan who sent a response directing Dallaire to do nothing about the cache. More crucially, he failed to bring the cable to the attention of the Security Council. And later that month, when Dallaire again sent a cable outlining his plan to seize the weapons, he received another cable from Annan in effect telling him not to move.

It is clear that Annan's actions in regard to Rwanda will continue to haunt him. It is a huge and terrible scar on his record, but I think the rest of his record must be taken into account. Kofi Annan was a diplomat par excellence who greatly enhanced both the role and the image of the United Nations throughout Africa and the world. He introduced important and lasting changes at the United Nations, and in doing so he not only helped shape the organization for the better, he made Africa proud. Liberia was pleased to honor him with one of its highest awards when he visited the country in July 2006.

When I was about to leave the United Nations to return to Liberia, I went to meet with Mr. Annan to announce my decision.

"Are you sure you want to do this?" he asked. "You have a great future at the UN."

"Thank you," I said. "But I am sure."

"If that's what your heart tells you to do, then go and do it," Kofi Annan said to me. "Go and do well and make it work."

# CHAPTER 13

# WAR SOME MORE/1997 ELECTIONS

IN OCTOBER 1992 the cease-fire that had brought a desperate but fragile peace to Liberia exploded. Charles Taylor, fortified and impatient to continue what he had started years before, launched an all-out campaign to take the city he coveted: Monrovia.

I was at UNDP at the time, still fairly new in my position and required to keep out of Liberia's internal politics. So I watched from a distance as war erupted once again. ECOMOG responded to Taylor's offensive by accepting the aid of other warring factions, by boosting its numbers in Monrovia, this time to nearly 12,000 men, and by bombing ports and areas under Taylor's control, including the Firestone rubber plantation.

When Taylor's troops abducted and brutally murdered five American nuns living outside Monrovia, the international community, which had drifted away from the conflict, rose again in condemnation. The U.N. Security Council, which to this point had addressed itself to the Liberian crisis primarily through a statement calling on all factions to honor the cease-fire, now moved to impose an arms embargo on Liberia. All factions save ECOMOG were subject to the embargo, as the United Nations attempted, in a somewhat feeble fashion, to halt the river of weapons flowing into the country. (Many years later, in July 2008, I honored each of the nuns posthumously with one of the nation's highest decorations.)

In addition to the arms embargo, ECOWAS imposed economic

sanctions on the territories under NPFL control. The result was an economic divide between Monrovia and the rest of the country, an area containing nearly all of the country's exploitable natural resources and the main source of food production. Contrary to the intended effect of the sanctions, the country's productive potential lay in the hands of the warlords, while Monrovia, the center of the nation's financial, commercial, and business services, slowly but steadily decayed.

Following repeated requests by ECOWAS countries for greater support in the ECOMOG intervention, the U.N. secretary-general appointed a special representative. The appointment of a special representative can be an important step for countries in crisis, ensuring a high-level advocate whose focus will remain on the issues at hand. That the United Nations had delayed so long in taking this critical step was disappointing to many. Likewise, the OAU soon appointed an "eminent person," finally signaling its official support of the ECOWAS initiative.

Critically, though, by this time yet another faction had entered the fray. Calling itself the United Liberation Movement of Liberia for Democracy (ULIMO) and composed of former soldiers and displaced persons who had fled to Sierra Leone and Guinea during the fighting in 1990, the new faction began asserting itself into an already crowded, ever-expanding field of combatants.

Soon afterward, the major rebel factions splintered into subgroups and rival ethnic factions, creating an alphabet soup of warring forces that looted, raped, and brutalized the land. Taylor's NPFL threw off several dissident groups with floating allegiances. A group called the Lofa Defense Force (LDF) rose in the wake of claims of ULIMO atrocities against non-Mandingo citizens of Lofa County. Another group, calling itself the Liberian Peace Council (LPC) and headed by George Boley, Doe's minister of presidential affairs after the coup d'état, emerged in the southeastern part of the countryside. The LPC engaged the NPFL in that area, sending some 40,000 citizens fleeing from their homes. Most landed as refugees in Côte d'Ivoire. ULIMO splintered into the Krahn-dominated ULIMO-J, headed by Roosevelt Johnson, and the Mandingo-focused ULIMO-K, headed by Alhaji Kromah.

Liberia fell into balkanized chaos. The dark years began again.

Increasingly, the warring factions filled out their ragtag armies with younger and younger "recruits." These were the years of our on-again, off-again civil war, fought, first and foremost, by our own children. These were the violent years that brutalized and made brutes of an entire generation of our young people.

No faction refrained from the practice of enlisting child soldiers, but Taylor, in particular, was a master of the game. He recruited unschooled, unsophisticated, wide-eyed country boys, some as young as age nine, many of them orphans whose parents had just as likely been killed by Taylor's own forces as those of his rivals.

Sometimes Taylor's commanders would force the boys to kill their own parents and elders. In traditional African society, where elders are revered and respected, this is the ultimate taboo. Once family ties and traditional values were sliced in half, the boys were easy fodder for "Pape" to step in and claim as his own. They were entranced by Taylor; they would follow him to the ends of the earth.

His commanders gave the recruits drugs—opium and marijuana and amphetamines—to fuel their ferociousness. They gave them amulets and potions and other tokens of magic to "protect" them from bullets and fuel their fearlessness. They gave them guns and the permission to take whatever they wanted, then sent them out to village and countryside to loot, rape, fight, and kill. They did so with frightening enthusiasm and efficiency.

Writing in the *New York Times,* the reporter Howard W. French told of how one young fighter, rifle at the ready, described for French the death of his family at the hands of NPFL troops. In the next breath he described his alignment with the force. "I joined them because they are the best," said the child.

French also wrote about how in one offensive against Monrovia, Taylor's child soldiers destroyed buildings at the university, marveling at the toilets they found, which they mistook for "white men's wells." This is very clearly the failure of development in years past.

During this time the political process stumbled along in fits and starts. Between 1990 and 1995, there were, by some accounts, as many as fifty separate peace talks or peace conferences, producing as many as a dozen distinct peace accords.

In July 1993, a peace agreement known as the Cotonou Accord called for the creation of a five-member Council of State to replace Sawyer's interim Government of National Unity. The council would govern until elections, scheduled for February 1994.

The agreement also called for an immediate cease-fire and for all warring factions to disarm and go into rehabilitation camps under the supervision of ECOMOG and a small United Nations Observer Mission in Liberia (UNOMIL). By October, that force, some three hundred men, had arrived in Liberia. But the splintering of the ULIMO militia and Taylor's refusal to disarm his troops doomed that accord.

Not until August 1995, after the United Nations warned it would end its observer role in Liberia if progress was not made, did the factions reach a more stable power-sharing agreement. During talks in Nigeria the six warring factions and a representative of the Liberian National Conference signed an accord that created a six-member Council of State.

Wilton Sankawulo, an English professor who had not been involved in the war, was chosen as chairman. Taylor was given a seat, as were Kromah, George Boley of the Liberia Peace Council militia, a civilian politician, Oscar Quiah, and a traditional chief, Tamba Tailor, who was in his nineties.

That September the second National Transitional Government of Liberia (NPFL) was seated in Monrovia. Huge crowds converged in the streets of Monrovia at a ceremony marking the transition, evidence of the Liberian people's desperate hope that peace was finally at hand. But hardly a month later new violence erupted between the NPFL and ULIMO.

By April 1996, Monrovia was once more caught in the cross fires of factional bickering, sparked by the dismissal of Roosevelt Johnson from the Council of State. The council lost all authority as Taylor and Kromah battled to see which one would take power of Liberia by force. Once again bodies lined the streets while civilians dodged and hid. Once again helicopters landed at the U.S. Embassy compound to evacuate Americans, leaving Liberian coworkers to stay and die in the swirling dust. The International Red Cross pulled out, citing "total anarchy" in the city. The U.N. High Commissioner for Refugees described the scenes as "wanton carnage, totally out of control." ECOMOG troops abandoned their checkpoints and retreated to their barracks to wait and see what the end would be. Between lulls in the fighting, tens of thousands of people roamed the streets, searching for water and food.

More than three thousand people lost their lives in this new round of fighting before it was brought under control by ECOMOG in August. As the nearly one million people crowded into Monrovia began, once again, to pick up their lives amid the rubble, ECOWAS met again in Nigeria to establish a new timeline for achieving peace.

The feeling this time was that Nigeria and the other ECOWAS countries were reaching the end of their patience with the Liberian situation. Nigeria in particular signaled an intensifying of its commitment to do whatever it took to hold the cease-fire long enough for elections to be held and the country turned back over to a democratic ruler.

Under the new agreement, gone as head of the council was Wilton Sankawulo, who was accused by some of backing Taylor and who was generally considered unequal to the task at hand. In his place was chosen Ruth Sando Perry, a widow and grandmother who declared herself "hard as steel" and determined to force the warlords into line.

Perry had not sought the position, and her choice was a surprise to many. To be honest, Ruth was not one of my favorite people at the time. She had been among those of us in the grand coalition group who had vowed not to take our seats after the disputed 1985 elections, but Ruth had broken ranks and taken her seat. She had become a senator and served during the Doe regime, and that was my impression of her at the time she assumed the council leadership.

It was clear the warlords had settled on her because she was known in Liberia but had, since the events of 1989, been a private and uninvolved citizen. No doubt they considered her perhaps more flexible than others who might be chosen and no threat to themselves. She was given no real authority, staff, or money with which to do her job. It was a surprise to many, then, when, using only the moral authority of her office and her own sense of right and wrong, she determinedly used her voice to scold the warlords into disarming their troops and submitting to the election process. She took little concrete action, but then again she was there in the midst of all the warring parties and had she tried to take any strong, independent action, they would no doubt have moved against her. The mere fact that she survived her term, given the situation, is a huge credit to her.

During the 1997 elections, although she remained publicly neutral, as was required by her position, Ruth was, behind the scenes, very supportive of our campaign. We became close friends during that period, and we remain so to this day. When she is home, we visit and talk together.

Elections were scheduled for May 1997. By January, the momentum was palpable; this time, it seemed, all the powers on the ground had gotten serious about ending the war and holding elections. Outstripping expectations, thousands of fighters from all sides lined up at designated centers across the country to surrender their weapons in compliance with the disarmament portion of the peace accord.

By March, the newest commander of ECOMOG was declaring the disarmament largely complete and the country safe for elections. No one really believed that all the weapons in Liberia had been turned in; ECOMOG forces uncovered arms caches in villages and homes on a regular basis long after the official disarmament was complete. Nonetheless, the progress was sufficient, and the hunger for peace and stability was so deep, that things moved ahead.

While all these events took place, I remained in New York at UNDP working for all of Africa. I had drifted away from the Liberian

Action Party and was not associated with any particular group. Still, in early 1997, I was called by some members of the LAP and asked to come home to help contest the elections. An alliance was being formed among six or seven political parties, including the LAP, and they wanted to choose a consensus candidate who would be strong enough to take on Taylor. "There is no one who can challenge Taylor as effectively as you can," they said to me.

I did not immediately agree. Life at UNDP was both challenging and fulfilling. I was making a very comfortable living, got to travel extensively, and had none of the pressures associated with governing in the best of times—not to mention trying to piece back together a war-torn land. A lot of my colleagues at UNDP thought I had gone cuckoo to even consider the idea. But I had always looked for challenges, and this was one of the greatest of all.

I decided I would meet with Gyude Bryant, who was then chairman of the LAP, to discuss the issue. As I was going to Dakar on business anyway, we agreed to meet there to talk things through. Gyude told me I needed to make a decision about whether I would seek the nomination or not, because things were moving quickly. We agreed on a date for the party's convention to choose its candidate, and I agreed to return home to attend the convention and told him that he would know my decision by then.

Back in New York, I worked through the pros and the cons of accepting the challenge and decided I wanted to run. The first thing I had to do was to tell my family, not all of whom, I knew, would be pleased by the idea. My sister and brother, who loved me very much and whom I deeply loved, were both private people who had never fully understood my desire to live a public servant's life. Jennie had told me more than once that with my education and training I could better help my country in the private sector, without all the hassles, uncertainties, and downright danger of being a politician in Liberia.

Around the time I was deciding to run for president, my family and I all went on a cruise to celebrate Jennie's sixtieth birthday. Later

Jennie told me she knew something was up when I agreed to go on the cruise, because I don't like the confinement of a cruise ship. It is not my type of relaxation or fun.

One afternoon when everyone was gathered, I made my announcement. "I have decided to run for president against Taylor," I said.

Well, they got upset. "You can't do this!" they said. "If you want to help the country, you can help from where you are, from the United Nations."

I had never in my life felt as isolated as I did that day on that ship. I felt as if my family were disconnecting from me and from all the things I had been working for all those long and difficult years. Not a single person agreed with my decision; not a single one said, "We support you."

They could not understand at all. They felt I was comfortable in the United Nations, I held a high position with all the exposure and contacts and potential for advancement I could possibly want or need. I had an apartment in New York just a few blocks from the United Nations, and it was clear I was on my way up in the organization. Jennie reminded me that Kofi Annan had suggested that I stood a good chance of being UNDP associate administrator one day.

"You can help your country that way," Jennie pleaded. "Don't do this."

No one could understand my decision to throw it all away to risk everything in a run for president. Particularly since I would be running against Charles Taylor, who had every advantage on his side— and if he didn't at first, he would soon make sure he did. It seemed to all my family just a crazy, stupid thing to do. They shook their heads and looked at me as if something was obviously not right inside. It was the one time in my life I felt my family was not with me, that I could not depend upon them, and it was very difficult.

Still, my mind was made up.

One day in late March, I received a message from a friend that the date for the convention had been moved up—suddenly and without notice to me. I immediately made plans to fly to Monrovia and arrived on the very day the convention was being held. I drove straight from the airport to the Centennial Pavilion, arriving just in time to learn that all the backroom wheeling and dealing was over and the die had been cast. Cletus Wortorson, a geologist, had been picked as the candidate.

What could I do? I went into the convention hall and said to the delegates, "What you will do today, you will feel consequences of it for years to come." Then I left and flew back to New York, thinking that was the end of that.

Soon afterward, however, I received another call, this one from Toga McIntosh. Toga was then with the Liberian Unification Party, part of the alliance. But the alliance was fracturing. Bacchus Matthews had pulled out his United People's Party and was running as a candidate. So was Togba Nah Tipoteh of the Liberia People's Party. The Unity Party, which had not backed a candidate in the alliance election but wanted to support whichever compromise candidate was chosen, had also withdrawn over concerns about the process. Toga McIntosh was deeply concerned about all the infighting and splintering and the impact such turmoil would have on the chances of winning the election. It was suspected too that Taylor had infiltrated the alliance and was throwing money around to keep the group so fractured that no single candidate would be able to contest him effectively.

"The truth is, none of these people is going anywhere," Toga McIntosh said to me. "Why don't we just try to get you to be a candidate?"

So once again I returned to Monrovia and began meeting with people, in particular the leaders of the Unity Party. Well, Monrovia is a small city and few events or movements go unnoticed for long. The town was full of newspapers and journalists, and somehow the BBC got wind of my deliberations. A reporter called to ask if he was correct in his understanding that I was to be a presidential candidate. I said, "I'm here and we're talking about it."

That was enough for the United Nations. Soon after the report aired, I received a telephone call from Kofi Annan's chief of staff.

"You're in Monrovia," he said.

"Yes."

"We understand you are there politicking. We understand you may be a candidate for president of Liberia. We wish to know your intentions in this regard."

At that stage I was already committed to the challenge. My heart was there, with my country. I could not turn back.

"I will be returning as soon as possible to New York," I told the chief of staff. "But, yes, please tell the secretary-general I will have to resign."

I flew back to New York, concluded all the things that needed to be concluded, and tendered my resignation. Then I returned to Liberia and began to campaign in earnest as the candidate of the Unity Party.

The election had been rescheduled for May but was postponed until July 19 to give the Elections Commission more time to prepare. Even that date was contentious—July is the month of the heaviest rains in Liberia. During the discussion on date setting, the United Nations and most of the country's political parties—with the notable exception of Charles Taylor's own—lobbied for delaying the vote until October, when the rainy season had passed. However, ECOWAS did not want to delay anymore. Nigeria and Ghana wanted their troops home as soon as possible, and ECOWAS had committed to keeping its forces there only six months after election day.

By June there were twelve declared candidates, including myself, Taylor, Cletus Wortorson, Baccus Matthews, Togba Nah Tipoteh, Chea Cheapoo of the PPP, George Boley of the National Democratic Party of Liberia, and Alhaji Kromah.

My decision to enter the race displeased a lot of people already on the ground, who felt I had come in and swept things away from them even as they were continuing to struggle to form a coalition. But it was clear that they were not going to succeed; Taylor had successfully splintered the alliance. Divide and conquer was a practice he employed fully.

Once my campaign got ramped up, most observers agreed that the real race would be between Charles Taylor and me.

We fought as hard as we could, waging a campaign of such energy and excitement it surprised many people. We had long, long marches and huge parades; our events were very lively and vigorous. The crowds we drew were quite large, sometimes stretching all the way from Broad Street to Sinkor. They cheered and chanted and sang their support, without fear of Charles Taylor, especially in Monrovia.

There were times when our supporters and Taylor supporters would encounter each other during a rally, and there would be taunting and jostling in the streets. I remember once our people standing up to some Taylor thugs who were calling me names. "What you know about Ellen, you dirty rouge?" they cried in return. "What you know?"

More than once in the middle of a rally we would look up to see Taylor flying over in his helicopter, trying to intimidate the crowd. This kind of behavior led some among us to fear that at some point Taylor would get desperate and try to eliminate the competition. I never worried about that possibility. The truth is, Charles Taylor knew he had the means to win that election, and he wanted the legitimization it would confer. After all those years as a rebel leader, he wanted to be recognized as more than that. All he had to do was to make sure he won the election, and he would then become the legitimate ruler of Liberia.

So our events were lively and well attended, and we carried much sway among the crowds. But how did all those people vote? We'll never know the answer to that important question. In retrospect, I believe it is likely that many of the people who came to our rallies did not, in fact, register to vote. Later we realized that many of the people, especially the women, did not know they had to register beforehand. When election day arrived and they were turned back from the polls, they were surprised and so were we. It was a good lesson to learn: make sure people get registered.

It was, of course, right in the middle of the rainy season, which made campaigning really constrained. Outside the capital, most of the roads in Liberia are dirt; in the rainy season they turn to mud, with large, gaping, water-filled holes that can swallow the front end of a car. Traversing them in anything less than a four-wheel-drive vehicle is impossible, and even in such a vehicle the going is infinitely slow, exceedingly bumpy, and terribly difficult. I got into the countryside as much as possible, even managing to pull together the funds for a few brief helicopter rides. But there was no way I could cover the whole country; I simply did not have the resources, and the weather did not permit.

Despite the challenges, I have always enjoyed campaigning, and I enjoyed that particular campaign just as much. I like getting out and talking to people, wading into the midst of crowds to hear what people have to say. I like too the more technical side of campaigning, the strategy and planning, the formulation of campaign slogans. Ours was simple: "Vote for change."

This was the election that drew President Jimmy Carter to Liberia as an observer. President Carter had a long-standing interest in the country, having first visited when he was president in 1978, the first time a sitting American president had visited sub-Saharan Africa. By the time of the 1997 elections, then former President Carter had established friendly relations with Charles Taylor. It is fair to say that President Carter had been charmed by Taylor, convinced by him that he was a good Baptist at heart. Many others, including Reverend Jesse Jackson, Pat Robertson, and Congressman Donald Payne, were likewise convinced enough to support Taylor.

In an article published in the *New Yorker*, Representative Payne said he liked Taylor because "he's intelligent; he knows what sells here, and he's from over there. He has the knowledge of both worlds."

During the campaign, President Carter and his wife, Rosalynn, visited me at the house I was staying at in Monrovia. President Carter asked me what I would do once Taylor won the elections, because he was fairly certain Taylor would win.

I said, "I'm not sure, Mr. President. We plan to give Taylor a run for it. We're not going to just bow out. We're going to compete with him."

President Carter said, "Yes, but I'm afraid the odds are just against all of you. Taylor is going to win."

I still believed we had a fighting chance to beat Taylor, and I told President Carter as much. Later, in a trip report to the Carter Center, he described our meeting this way: "She was critical of Charles Taylor, stating that she would not accept his victory or serve in any way in a government if he should be elected. Under our questioning about accepting the results of an honest election, she only became more vehement."

If I was vehement, it was because I knew full well the election would not, in fact, be honest. In the end, President Carter was right about this much: Charles Taylor was going to win the election, because Charles Taylor had several critical advantages.

The first was money: a seemingly bottomless supply of funds he had accumulated by looting the country during seven years of brutal war. Taylor mobilized a fleet of motorboats, motorcycles, and loudspeaker trucks. He flew around the country in a twelve-seat helicopter, buzzing the rallies of his opponents and handing out bags of rice, pens, color posters and stickers bearing his image, and thousands upon thousands of T-shirts to eager crowds. The T-shirts were wildly popular.

Taylor also still exercised strong control over much of rural Liberia. He controlled several newspapers and two radio stations, including the only station capable of broadcasting beyond Monrovia. In a country where much of the population is illiterate and radio is the prime instrument of communication, this was a powerful advantage, and Taylor made full use of it. Some reports suggested that people in very distant villages had not even heard of the other candidates.

More than anything, Taylor had one huge rock on his side that was more than sufficient to tip the scales his way: fear.

There was a notorious song popular at the time, a song Taylor's former child soldiers and future voters serenaded him with as he campaigned across the countryside. The lyrics went like this: "You kill my ma. You kill my pa. I will vote for you."

The message from Taylor—hinted at, implied, crystal clear—was this: vote for me, or you'll regret it. Many Liberians felt that a vote for Charles Taylor was a vote against returning to war and vice versa. They feared that if Taylor did not win control of the country by the ballot, he would resume his attempt to take it by force, thus immersing the people in another bloodbath. In that regard, he had overwhelming support. There was no doubt about that.

Another of Taylor's campaign slogans was "Better the devil you know than the angel you do not." That is what many people believed.

During the course of the campaign many people said to me, "I believe you are the better candidate, and I believe you would be the better leader. But we just can't take that chance. We want peace, and this is the only way we'll get it."

"Mr. Taylor has fought hard to get this country," the people said. "This is his opportunity to get it."

I went to Ghana, Côte d'Ivoire, Nigeria, and other countries to try to generate regional attention, but there was no interest. Only leaders from outside the ECOWAS zone were curious and courteous.

Of course, Taylor also resorted to the negative. He declared that no woman could govern Liberia. Who had ever heard of such a thing in Africa? He told the voters that he had sacrificed long and hard, had suffered and fought to bring "freedom" to Liberia, and now those he viewed as interlopers were trying to snatch from him the earned results of that sacrifice. Never, he declared. Not over his dead body. People knew what that meant.

On election day the Nigerian minister of foreign affairs, Tom Ikimi, with whom I had had a good relationship from my days at UNDP, escorted Taylor to the polls—a clear demonstration of Nigeria's support for Taylor's candidacy, despite the fact that Nigeria was a major player in planning and monitoring the elections.

In the end, Taylor was declared the winner, with more than 75 percent of the vote. Although before the elections many observers had expected no one to receive a clear majority and a possible run-off between Taylor and myself, in the end I was merely declared the runner-up, with 9.5 percent—pointedly just under double digits. We thought that was cute.

Taylor's NPFL also won the overwhelming majority of legislative seats, giving him complete and utter power.

When the results were announced in late July 1997, we did not challenge them, knowing that to do so would be a waste of time. I did decry the fraud that I felt—and still feel—took place. Taylor may have won, but not by the numbers that were claimed.

However, more than three hundred international observers, Carter included, declared the election free and transparent, if not precisely fair. Carter went so far as to declare the process and resulting transformation of Liberia from war-torn land to what he suggested was a functioning democracy "almost a miracle."

In an interview with the filmmaker Nancee Oku Bright, Carter added, "I can't say that it's been completely fair, and neither are the elections fair in my country, where if you don't have $50 million to spend, your voice will never be heard, through television or radio, by the people of America. So you can't say that it's completely fair unless all thirteen candidates have the same amount of money, the same number of vehicles, the same amount of access to news media, and that has not been the case. But I think as far as this election representing the freely expressed will of the Liberian people, I would say yes, it has been a free election.

"I don't think there could have been any fraudulent actions without [their] being detected, and all of the candidates have pledged to support the results of their elections, whether they win or lose. You can't say that about many elections on Earth. So I'm very proud of the Liberian people."

Taylor called me following his victory, but I refused to take the call. Later, President Carter came to visit me again and to ask about my plans. He wanted to know if I would work with Taylor in his new government.

"I think it would be a good thing," the president said to me.

I was astonished. There was no way in the world I would work with Charles Taylor after all the death and destruction he had caused, and I told President Carter as much in no uncertain terms. What's more, I still believed Taylor had engaged in fraudulent elections, regardless of the stamp of approval the process had been given.

"Do you really think I'm going to identify myself with that?" I asked Mr. Carter. "No, President Carter."

I told him he should be speaking out against the fraud in the elections instead of trying to cajole me into joining Taylor's government.

I think the president got a little miffed with me because of that. He left our meeting soon afterward, clearly unhappy. In his report to the Carter Center on the trip he suggested that I was unreasonable, a poor loser, and not a team player. That was that. It took many years to mend that rift, especially once the conversation was picked up and spread around, as conversations of that magnitude will be. It was said that I had not spoken kindly to President Carter nor received him warmly. But it wasn't the messenger I was responding to, it was the message. A message I could not accept.

At any rate, Taylor had won. He was president of Liberia, and his victory was accepted by the Liberian people and the international community. He finally had what he wanted. The question was, what would he do with it?

# CHAPTER 14

# SELF-IMPOSED EXILE,
# OR EXILE AGAIN

AT HIS inauguration, Charles Taylor promised, among other things, "not to be a wicked president."

I was not in attendance. I had left the country, flying back to New York to wrap up some things and plan the next stage of my life. On one of my several return visits I did have a meeting with Taylor and some of his cabinet ministers. I tried to work with Taylor, as I had done with Samuel Doe, to offer my experience and knowledge, to suggest some of the things his administration needed to do in order to lift the country back onto its feet, economically and socially. The meeting was a waste of time; Taylor would not listen. He was a man who always thought he knew everything he needed to know about every subject. He lectured about economics. He boasted about his knowledge of the American political system. There was nothing he could learn from you, and he viewed any challenge to his knowledge as a threat. Earnest Eastman was in the meeting as his foreign policy advisor, the same Earnest Eastman who had sat in a similar meeting with Doe years before. Realizing there would be no compromise, I thanked Mr. Taylor for his time and went my way.

Early after his inauguration, Taylor sent me one of his longtime associates from his GSA days by the name of Blamoh Nelson and asked me to join his government by leading the social security agency. My response to Blamoh was "Great—he wants me to take care of the people while he steals the money? Thanks, but no thanks."

Blamoh, who was a staunch supporter of both Charles Taylor and Baccus Matthews, whom Taylor despised, is today a senator.

With my tenure at the United Nations at an end, there was no need to move back to the United States. Moreover, I wanted to be close enough to the scene in Liberia to monitor it and to work on it if events progressed as many of us feared they would. I decided to move to Abidjan, where I set up a small financial consultancy firm with the help of my colleagues and connections from Equator Bank. It was from that vantage point that I watched, along with the rest of the world, to see what Charles Taylor would do with the prize he had fought so long and so hard, and sacrificed so many lives, to attain.

The answer did not take long to arrive.

Right from the start Taylor signaled his unwillingness to tolerate criticism or opposition by shuttering radio stations, closing several newspapers, and arresting journalists, some of whom were beaten or whose homes or offices were ransacked and destroyed.

He appointed some outsiders to key positions but filled many others with his key militia chiefs. He began filling the ranks of the police, national security, and the Armed Forces of Liberia with his former soldiers, turning them all into his own personal armies. The Abuja Accord had assigned the task of restructuring the AFL to ECOMOG, but after his election Taylor demanded the right and largely took over the job, booting out ECOMOG, retiring thousands of former army people, many of them Krahn, and replacing them with his loyalists. In essence, Taylor was rearming himself, knowing that his methods would eventually begin to be contested.

In the meantime, the entire Liberian economy was slowly being taken over by criminals and cronies. Drug rings were widely suspected to be operating with impunity. Taylor handed out large concessions to friends and shady businessmen, people who entered the country in collusion with government officials and benefited from its precious resources, with little benefit to the country as a whole. Notable among this was the forestry operation granted the Oriental Timber Company of Malaysia, which reportedly amassed riches from what Taylor called his "pepper bush." According to the *San Francisco Chronicle*, a 2001 report by the London-based watchdog organization Global Witness accused Oriental Timber of clear-cutting forests, exporting undersized logs, and cutting an unprecedented 131,000 cubic yards of forest in its first twelve weeks of operations in 1999. The company's chairman, Gus van Kouwenhoven, a close associate of Taylor, was convicted in 2006 of selling arms to the Taylor government in violation of a U.N. embargo, but an appeals court later overturned the conviction. Kouwenhoven was also on the board of the Liberian Forest Development Authority, whose managing director was Taylor's brother D. Robert Taylor.

At first the opposition, both inside and outside the country, tried to mount pressure on Taylor to follow the rule of law. We met with Liberian and international groups, contacted human rights organizations, and wrote articles for newspapers. Every now and then I would slip back into Liberia, make public statements about what was going on, then leave. This behavior irritated Taylor, of course. Soon it was very clear that he and I were on a collision course, with neither of us knowing where it would lead.

Sometimes stories would drift back to me, rumors of Taylor planning to send assassins to Abidjan or to attempt to hit me when I came back into Monrovia. I never worried about it. Sometimes people find this surprising about me, but I believe that perhaps the thing that has sustained me during my many close calls is that I have a certain frame of mind that banishes fear.

I suppose that's the part of me that believes in predestination. I believe that when it's my time to go, it's my time to go, and when that time arrives—so be it. Nothing I do or do not do will be able to stop it. Until then, however, I plan to keep doing what it is I have to do.

In December 1997, Samuel Dokie, a former colleague of Taylor's who had broken with him and run for the Senate on the Unity ticket, was abducted and brutally murdered, along with his wife and two other relatives. The Dokies were on their way to a family wedding in

Nimba County when they were arrested by police in Gbarnga, Taylor's stronghold. Their burned and mutilated bodies were later discovered in their cars. The police claimed they had been acting on the orders of Benjamin Yeaten, Taylor's chief bodyguard. Taylor denied any knowledge of or responsibility for the deaths.

Yeaten, the executor of many an order by Taylor, today remains at large somewhere in the West Africa subregion, plotting and hoping for Taylor's return.

The deaths of the Dokie family sparked mass demonstrations in Monrovia and dashed hopes that Taylor would somehow rise above his warlord tendencies to true leadership. Likewise the death of Norwai Flomo, a persistent critic of Taylor, who was dragged from her home by state security forces and killed. To many of us these atrocities and criminal acts confirmed that Taylor's motives in seeking the presidency had been personal aggrandizement, individual empowerment, and financial gain.

From Abidjan, we mounted an opposition and began working to sensitize the international community in general, and the leadership of Africa in particular, to the fact that Taylor was bad news. He was not the person for Liberia or for the region, and he would lead us to disaster—that much was clear. At the same time, others were also fighting to raise the alarm about what was happening in Liberia. People such as Amos Sawyer in the United States and others were lobbying and mobilizing.

Unfortunately, despite our efforts, through much of 1998 and into 1999, many people continued to give Taylor the benefit of the doubt. Even the Commonwealth of Massachusetts dismissed the still-standing charges against him, saying the U.S. State Department had asked as much "in the interest of harmonious relations between the United States and Liberia." The dismissal was an important symbolic victory for Taylor. Not only did it mean he was free to visit the United States, it also signaled that the international community stood ready and eager to welcome him in.

He never did make that much-desired trip to the United States, though—perhaps out of fear of what the Liberian people would do in his absence.

THINGS CHANGE, however. By the dawn of the new century, the world was waking up to the truth of Mr. Taylor.

This happened not simply because of Taylor's human rights abuses, blatant looting of national resources, and decimation of the economy at home but because, emboldened by success, he had decided to branch out. He could not contain his adventurism; he coveted the rich resources of neighboring countries, and that drove him on.

In March 1991, a small group of armed men calling itself the Revolutionary United Front (RUF) began a civil war in eastern Sierra Leone. Taylor, who by this time controlled much of rural Liberia, gave the group weapons and territory from which to attack the forces of Joseph Saidu Momoh, Sierra Leone's then president.

Led by Fodoy Sankoh, a former officer in the Sierra Leonean army, the RUF became a brutal and deadly force that quickly gained control of Sierra Leone's valuable diamond mines and plunged the country into nine years of civil war. Taking a page from Taylor's book, the RUF recruited child soldiers by forcing them to rape or kill their own parents before swearing allegiance to Mr. Sankoh. These armed youths then operated not just by killing their opponents but by terrorizing tens of thousands of innocent civilians caught in their cross fire. The group's signature tactic was to hack off the hands, feet, arms, or legs of their victims. Men, women, children, it didn't matter. After one has raped and killed one's own mother, what boundaries are left to cross?

All of this took place with Taylor's direct encouragement and assistance, both financial and military. Taylor's support for the astonishingly violent RUF was the single biggest factor in the ability of that group to continue waging Sierra Leone's devastating civil war. That war cost tens of thousands of people their lives, left tens of thousands brutally mutilated, and displaced nearly two million.

By mid-2000, Taylor's involvement in this nightmare was increasingly clear to all concerned. Even Jimmy Carter had lost faith. In November of that year, Carter closed down his Carter Center in Monrovia. His staff was being intimidated and harassed in the wake of Carter's increasingly public criticism of Taylor for human rights abuses at home and relentless and destructive warmongering abroad.

In a public letter to Taylor President Carter wrote:

**❝** I AM very disappointed about the course of events in Liberia over the last three years, especially given the hopeful opportunities that were present after your election in 1997 following a terrible seven-year civil war.

For a period following your election, you and I maintained a dialogue in which I repeatedly offered to assist you in efforts to demonstrate your commitment to building a genuine democracy in Liberia. On several occasions I raised serious concerns about developments in Liberia and unsuccessfully suggested specific actions your government could take to address problems. Unfortunately, however, the dialogue seems to have broken down, and it has become clear that your government does not share the same goals.

Much to our dismay, Liberia is a country where reports of serious human rights abuses are common, where journalists, human rights organizations, and political activists work in an atmosphere of fear and intimidation, and where there is little political space for meaningful democratic debate. Instead of being used to improve education, infrastructure, and development, Liberia's resources have been diverted toward extra-budgetary uses. In addition, it is increasingly evident that Liberia's role in the conflicts of the sub-region has been a destructive one. **❞**

Even while Taylor was busy exporting terror into neighboring countries, he had issues at home. By early 2000, a new insurgency group had arisen, this one calling itself Liberians United for Reconciliation and Democracy, or LURD.

LURD was a loose coalition of dissident groups, many of them formed in exile and all of them opposed to Taylor. The militia was led by a man named Sekou Damante Conneh and included former members of ULIMO-K and ULIMO-J under the leadership of Roosevelt Johnson, among other fragments.

LURD received early support from Guinea, whose president, Lansana Conté, harbored an implacable dislike of Taylor that stretched back at least to 1992. That year, at the ECOWAS summit in Dakar, Conté had put forth a motion to treat the NPFL as a terrorist organization, thereby changing the ECOMOG mandate from peacekeeping to peacemaking. During that meeting, he famously called Taylor a "cancer" that, unless cut out, would inflame the entire subregion.

Furthermore, by 2000 Taylor had already tried some incursions into Guinea. Conté, seeing this, and seeing Taylor's naked desire for Sierra Leone's diamonds, worried that it was only a matter of time before Taylor also began lusting after valuable national resources in Guinea. So he countered by supporting the LURD rebels.

LURD began attacking Taylor forces in Lofa County, battling its way south toward Gbarnga and Monrovia.

Sometime in 2000, Taylor tried to link me with the LURD dissidents and vowed to personally arrest me if I set foot in Liberia. I fought back to defend my name, releasing a statement that read, in part, "Mr. Taylor has indicated, as partial evidence of the accusations, that I was recently in Guinea. I defy him to expose details of this visit, to expose to the nation the date when I was in Guinea. If he is a true Darkpannah, let him reveal this little bit of what he calls 'fact.' This will be my opportunity to prove my innocence thereby exposing him as a pathological liar to use the characterization of him by a revered Liberian cleric some time ago."

(The cleric in question was Archbishop Michael K. Francis. Today, sadly, he is paralyzed but during those terrible years he was unrelenting in his condemnation of Charles Taylor. I do believe the very intensity he brought to battling Taylor perhaps contributed to the stroke he eventually suffered. During my inaugural speech I referred to this brave man as the conscience of the nation.)

But I could not deny that Taylor had the power to carry out his threat of arresting me. And since it was abundantly clear by then that Taylor had no respect for the rule of law and would not be bound by any accepted standards of evidence or justice, I decided to stay away from Liberia for a while.

I was, however, still mobilizing the civil opposition from my home in Abidjan. One of the first tangible signs of success came in May 2001, when the U.N. Security Council imposed sanctions on Liberia. The United Nations accused Taylor of fostering civil war in Sierra Leone by backing the RUF and running a guns-for-diamonds operation. The sanctions banned the import of Liberian diamonds and imposed travel restrictions on senior Liberian officials, including Taylor.

The sanctions were aimed at weakening Taylor financially, which they did. Unfortunately, the restrictions were not applied (until May 2003) to Liberia's valuable timber, which allowed Taylor to continue selling concessions and using the money to buy arms and ammunition and to pay his commanders and assorted private mercenaries in his employ.

Meanwhile, I continued concentrating my efforts on the leadership of Africa, many of whom I knew well from my days at Equator Bank and the United Nations. In particular I targeted Blaise Compaoré, the president of Burkina Faso. Both Burkina Faso and Libya had provided Taylor with significant military aid during the long civil war and Taylor, Compaoré and Libyan leader Muammar al-Gaddafi were all known to have a very close relationship.

However, when I was head of the Africa bureau I had been to Burkina Faso several times on development issues and had had the opportunity to meet President Compaoré myself. I knew him and he knew me, which made it easy for me to target him for a turnaround. I kept working on him, trying to illuminate the truth about what Taylor was doing in Liberia and how destructive it was. Finally, President Compaoré decided he would host a reconciliation conference in December 2001.

By this time I had met George Soros, the Hungarian-born businessman and philanthropist. We met at a meeting in Paris. Mr. Soros was close friends with Alassane Ouattara, the former prime minister of Côte d'Ivoire and a former economist for the IMF. Mr. Soros asked Mr. Ouattara to put together a group of African leaders for a meeting to discuss an initiative he wanted to launch in West Africa. The agency would be similar to ones that already existed in Eastern Europe and southern Africa.

That group became the Open Society Initiative for West Africa, an initiative dedicated to supporting and developing societies marked by democracy, good governance, the rule of law, basic freedoms, and widespread civic participation in fifteen countries in the region. As it says in the OSIWA mission statement:

66 OSIWA believes that it best serves by sustaining catalytic and innovative initiatives that add value to the efforts of West Africa's civil society. OSIWA seeks to collaborate with advocacy groups, like-minded foundations, governments and donors. OSIWA further recognizes the importance of incorporating global developments in building open societies and seeks a greater commitment to the region by rich nations. 99

As chairwoman of OSIWA I had many opportunities to connect with Mr. Soros. So when we needed funds for a reconciliation conference, I was able to turn to him and his close associate Aryeh Neier, who leads the Open Society Institute.

The conference was held in Ouagadougou in December 2001. I was there, along with Amos Sawyer and several others, and President Compaoré invited Taylor as well. His thinking was that if Taylor came, it could be an opportunity for true reconciliation. Those of us who opposed Taylor would give up our opposition and accept his right to rule Liberia. In return, Taylor would listen to our complaints about his actions and finally see that he needed to cease behaving like a warmongering commando and began acting like the democratically elected leader of a free society. We would all of us sit down and talk, resolve our differences, and achieve some form of unity.

Taylor declined either to attend the conference or to send any representatives. Instead he sent a message. President Compaoré, when he met with the assembled Liberians from at home and abroad, relayed it to us. It was this:

"I have God and I have guns," Taylor said. "I don't need anybody else."

That was it for President Compaoré, I do believe. He gave up on Taylor right then and there.

Other West African leaders had already turned. They could see for themselves—by the increasing numbers of refugees fleeing across their borders, by the skirmishes being supported and encouraged in their own backyards or just across the fence—that Taylor had to go. If he remained, the entire subregion would be plunged even deeper into turmoil, chaos, and war. And everyone was sick of war. Everyone except the warmongers.

After the conference Taylor once again tried to crack down on the Liberian opposition leaders who had attended and were still living in Liberia. By now, though, Taylor was experiencing difficulty on many fronts.

LURD was battling ever closer to Monrovia. By February 2002, the rebel group reached within twenty-five miles of the capital. Taylor, in his desperation, smuggled more and more massive arms into the country and declared a state of emergency. Amnesty International accused Taylor of using the declaration as an excuse to cover up his human rights violations, including the rounding up, jailing, and forced conscription of young men.

It was a terrible time. People were running helter-skelter, fleeing both LURD forces and the responding government troops. Monrovia was still desperately overcrowded and underresourced from previous conflicts; now thousands more people poured in seeking shelter and safety where very little was to be found.

Yet even as the grip tightened around his neck, Taylor continued trying to charm his way out of the situation. He flew to Morocco to meet with presidents Ahmad Tejan Kabbah of Sierra Leone and Lansana Conté of Guinea and agreed to end interference in their internal affairs. When the Unity Party planned a rally around my scheduled return to the capital in May 2002, Taylor banned all public gatherings except those approved by the government but staged a meeting with me and other officials in which he mouthed various platitudes about democracy and freedom of speech.

Meanwhile the rebels marched. In early 2003, another group, this one called Movement for Democracy in Liberia or MODEL, emerged in the southeast near Côte d'Ivoire and began its own assault on Taylor and the capital. MODEL was made up mostly of members of the Krahn tribe, soldiers who were anxious to avenge the death of President Doe. They obtained support from the government of Côte d'Ivoire, which charged Charles Taylor with interference in its own, nascent internal conflict.

The citizens of Monrovia were trapped in a desperate vise, squeezed between LURD in the north and west, MODEL in the east, and Taylor in their midst.

By June, the city was desperate. Rebel mortars pounded crowded neighborhoods, killing hundreds by the week and panicking those who remained. Machine-gun fire echoed day and night. The United Nations warned of a major humanitarian crisis as more than a million people remained trapped in Monrovia, cut off by the fighting from food, water, and medical supplies.

In late June, on the eve of his first-ever visit to Africa, President George W. Bush called on Taylor to resign for the sake of the peace. In early July, President Bush spoke more clearly and more forcefully. "Mr. Taylor," he said, "has got to leave."

We were overjoyed! Finally, there it was. Finally, a leader of consequence had spoken, a leader who had the power to back up what he said. Bush, with the weight of the United States behind him, signaled that Taylor had to leave and leave immediately. No more negotiations. No possibility, as Taylor had suggested, of serving out his term or standing again for reelection or yet another round of peace negotiations.

Taylor had got to leave. It was music to our ears.

Finally events were moving, though the nightmare was far from over. Secretary-General Annan appointed Jacques Klein, former head of the U.N. mission in Bosnia and Herzegovina, as his special envoy for Liberia. Klein was a tough-talking, battle-hardened military man, a retired major general of the U.S. Air Force and senior diplomat who labeled Taylor a "psychopathic killer" and warned him to get out. His job was to lead and coordinate U.N. activities in Liberia and to help stabilize the country ASAP, and he would fulfill that role.

Klein came to earn the respect of Liberians for his no-nonsense ways and his unequivocal refusal to play favorites among the warring factions. When skirmishes in Nimba between former government soldiers and MODEL endangered both civilians and the peace agreement in November, Klein sent patrols to fly over the area and threatened, essentially, to shoot anyone caught warmongering. The skirmishes stopped.

After Klein's appointment, the Security Council authorized the establishment of a multinational force to support a cease-fire in the area. The force would once again be led by Nigerian troops and those from other ECOWAS countries but this time would be supported from the onset by the United Nations and by major international players. The Bush administration pledged $10 million to help cover some costs and sent 2,300 marines aboard three vessels to idle near Liberia.

But President Bush refused to send troops actually onto Liberian soil until Taylor left the country. Taylor refused to leave until international troops arrived. It was a standoff, and it came at a terrible price for Liberians.

If June 2003 was miserable in Monrovia, July was worse. The rebels, sensing Taylor's weakening position, pressed harder, pounding the city with mortar shells. Taylor's forces responded in kind, leaving hundreds of thousands caught in the cross fire. In one week alone in July, more than six hundred people were slaughtered by mortar fire and machine guns. Monrovia was dubbed "the world's most dangerous place" by the international press.

As the streets filled with bodies, some ten thousand people sought shelter close to the U.S. Embassy residential compound. But amid such carnage, not even U.S. property was safe. One day a shell smashed into the compound, killing nearly thirty people instantly and wounding some three hundred more.

Terrified citizens hauled their loved ones to a ragtag Red Cross clinic in wheelbarrows. Others, wailing and screaming in grief, gathered the mangled bodies of their family members and lined them up in front of the embassy as a plea to the United States to send a peacekeeping force immediately.

The bodies lay before the embassy in the pouring rain all day and into the next, until finally they were taken up and buried by family members. Most were laid to rest on a nearby beach, out of necessity. The fighting made it impossible for family members to carry their dead to the cemeteries.

# CHAPTER 15

# ACCRA AND THE TRANSITION

"IF PRESIDENT Taylor removes himself for the Liberians, will that bring peace?" Charles Taylor asked, perhaps rhetorically. "If so, I will remove myself. If I am the problem and seem to stand in the way for Liberia to achieve peace, I will remove myself from the process to allow peace to come to our country."

Thus Taylor spoke during the opening session of the Liberian peace talks held in Accra, Ghana, in early June 2003. Like most people in the audience, I was a bit taken aback by hearing Taylor actually offer to give up what he had fought so ruthlessly to obtain.

I was also a bit skeptical.

Just days before, Taylor had sounded more like his usual defiant self. Upon arriving in Ghana to attend the ECOWAS-sponsored talks, Taylor said, "I was elected by eighty percent of Liberians, and the other twenty percent took to the bush. Therefore, I would not entertain any thought or even the possibility of resigning as president."

For good measure he added, "I am the president, and I shall remain president."

But even as Taylor raised his fist in defiance, it was clear to everyone else—to the United Nations and, increasingly, the United States, to the leaders of ECOWAS, and most of all to the people of Liberia—that he had to go. Monrovia was under siege, and hundreds were dying by the week. Thousands more were streaming from the country, seeking refuge and overburdening neighboring states. (Hundreds of Liberian refugees, particularly women, showed up at the conference, wearing T-shirts and waving signs calling for peace.) And as criminals and mercenaries from all corners of the continent got drawn into the fray, it was clear the Liberian conflict now threatened the stability and security not only of the subregion but of all of Africa. President Oluse-gun Obasanjo of Nigeria, speaking for a weary region, put it this way: "West Africans have made enough sacrifices for Liberians, so they must give peace a chance. . . . We have the will, we have the capacity, we have the ability, we African leaders, to solve African problems if our friends will just help us not to make war in Africa."

President Bush had not yet spoken, but Africa and the world had. The United Nations had renewed sanctions, adding timber to the list of banned exports. An impressive number of African leaders, includ-ing Thabo Mbeki of South Africa and Obasanjo of Nigeria, were all present or represented at the meeting and said to be in private agree-ment that it was time for Mr. Taylor to leave. The game was over. It was clear, however, that they wanted to give him, if they could, an easy, graceful exit from the stage.

What people who have never met Taylor personally may not fully comprehend is how charismatic he is—charismatic and charming and, to some people, genuinely likable. Many people did indeed like Taylor, including several high members of the African leadership. When he began to speak, when he cajoled and justified and explained away his terrible, terrible deeds, people had a tendency to fall under his spell. Many of the African leaders believed him when he said that his op-ponents were being unfair to him, that most of the trouble engulfing Liberia was caused by people who wanted to deny him what he had earned by both the ballot and the bullet: the right to succeed as the leader of Liberia.

What is more, the issue of noninterference in sovereign states still hung heavily over the heads of the African leadership. In their minds, Charles Taylor was the duly elected president of an indepen-dent nation. Yes, they believed, the election may have been tainted, but then elections in Africa often are. If you wanted to, you could credibly challenge so many of them. For the leaders, Taylor had been demo-cratically elected. To go after him, regardless of his misdeeds, would be establishing a marked precedent, which they were not completely sure they wanted to do.

That's why they wanted to give him a soft landing.

So the Ghana peace talks were arranged. The opening session would be held in Accra, after which the talks would shift to Akosombo, a city about 60 miles to the east. Invited to participate in the conference were Taylor himself, representatives of LURD and MODEL, eigh-teen of Liberia's political parties (including my own, the Unity Party), and five civil society groups such as the Inter-Religious Council of Li-beria and the Mano River Union Women's Peace Network.

On the first day of the talks, more than one thousand of us Li-berians gathered in the conference hall, waiting for the session to begin. It seemed to be taking a long time. There was movement at the podium, people hurrying around and across the stage, people coming and going from some back room somewhere. But the meeting did not begin.

Finally word filtered out and began rippling across the conference hall: Taylor had been indicted by the Special Court for Sierra Leone.

David Crane, an American and the court's chief prosecutor, said that Taylor was indicted for "bearing the greatest responsibility" for war crimes, crimes against humanity and serious violations of interna-tional humanitarian law in Sierra Leone since November 30, 1996.

"My office was given an international mandate by the United Nations and Sierra Leone to follow the evidence impartially wherever it leads," Crane said in a statement. "It has led us unequivocally to Taylor."

The indictment did not come as a total surprise. The court, a spe-cial partnership of the United Nations and the Sierra Leonean govern-ment, had already indicted nine people, including RUF leader Fodoy Sankoh, for war crimes and crimes against humanity in connection with the Sierra Leonean civil war. There had been persistent rumors that the court also sought Taylor's indictment.

Later we would learn that the indictment had been issued in March but released only in June. The timing took many by surprise, including the president of Ghana, who was not pleased.

In the conference hall gasps went up. Taylor had brought a forty-person delegation with him to the talks, and some of his people burst into tears, weeping and sobbing.

The rest of us were fairly pleased.

After a while Taylor came out and stood at the podium to address the crowd. Speaking of himself in the third person, he declared himself a sacrificial lamb and said he was willing to surrender so that Liberia might live. "If President Taylor is seen as a problem, then I will remove myself," he said. "I'm doing this because I'm tired of the people dying. I can no longer see this genocide in Liberia."

The great hall rippled with murmurs. Some people applauded this announcement of retirement, but the only people present who really believed what Taylor was saying were his own people. The rest of us were skeptical. How many times before had Taylor been on the ropes and escaped? How many times had he said one thing and done the opposite?

After Taylor's speech, he was whisked away by plane back to Monrovia. The Ghanaian government had been served arrest papers for him by the Special Court but declined to arrest Taylor, fearing the possible repercussions. It was said that Presidents Thabo Mbeki of South Africa, Olusegun Obasanjo of Nigeria, and especially John Kufuor of Ghana felt sandbagged by the release of the indictment just as their intense lobbying of Taylor had finally convinced him to resign—or at least to say he intended to do so.

"Obviously, this is an embarrassing incident, but as far as I am concerned, the focus should not be on our embarrassment," Ghanaian Foreign Minister Nana Akufo-Addo told a reporter for the *Washington Post* at the Accra airport as Taylor's flight departed. "I believe the action of the prosecutor in unsealing the indictment at this particular moment has not been helpful to the peace process."

In Monrovia, announcement of the indictment set off panic. Shops, banks, and businesses hurriedly closed, parents rushed to pick up children, people got off the street. People feared that if Taylor returned home there would be bloodletting. They feared that Taylor, seeing his empire crumble, would go after all his enemies, perceived and real. And they feared those enemies would retaliate.

In fact, Taylor's return did spark renewed fighting in Monrovia. The LURD rebels attacked the city after giving Taylor seventy-two hours to step down, even as their leaders continued to attend the peace talks in Akosombo. President Obasanjo flew to Monrovia to deliver in person an offer of asylum in Nigeria. Taylor accepted and promised to resign but then kept delaying his departure. Jacques Klein warned him to leave before he was arrested. "The warrant never goes away, and the court will be there for a number of years," Klein said. "Go while the getting is good." But August came and the fighting continued, and people continued to die.

During a break in the conference, I flew to Washington to lobby for the United States to become more actively involved in the conflict. Everyone was perplexed by the weakness of the U.S. response. I did what I could, then returned to Accra to keep working.

Then on August 4, 2003, the peacekeepers came to Liberia.

Hundreds of jubilant Liberians greeted the first wave of heavily armed Nigerian peacekeepers when they landed at Roberts Field. A few days later, thousands more lined the road into the city as the troops passed by in their caravans. The troops waved and blew kisses; the people waved and blew kisses back. "We want peace!" people chanted. "No more war!"

Meanwhile, in Ghana, we continued working on the peace talks as hard as we could. Once it was decided—at least in theory—that Taylor would step down, the natural question became who, or what, would take his place. Liberia would need yet another interim government, which would oversee the beginnings of reconstruction and reform and clear the way for free and fair elections of a new president and legislature.

It was slow, frustrating work. The talks, under the direction of an unseasoned ECOWAS mediation team, were unstructured, and the process was somewhat flawed. Meetings took place haphazardly, in ad hoc groups, with only the occasional plenary. The United States kept downgrading its diplomatic presence until only a relatively junior official remained.

There was early agreement on the security aspects of the plan, as well as on the mandate and tenure of the transitional government. Things bogged down, however, when it came to reaching agreement on the government's structure and composition.

Those of us in the political parties and civil society wanted a structure that would come as close as possible to what existed in the Constitution, with a single president and vice president. Taylor's representatives wanted mostly a continuation of the current structure, minus Taylor himself, while the various warring parties sought high-level decision-making roles for themselves and their representatives. This was the real sticking point, one that caused an early preliminary agreement in July to fall apart. The rebel representatives demanded that a power-sharing government be created, with three vice presidential slots designated for LURD, MODEL, and the Taylor government, a situation similar to what was happening in the Democratic Republic of the Congo. With the shadow of Taylor still hanging over the negotiations, the talks dragged and stumbled along.

Then, on August 11, the day Liberians had been waiting and praying for arrived. Charles Taylor, pushed by African leaders, by the United States, by the increasing chaos all around, finally resigned.

It was a surreal end to a nightmarish fourteen years. During a ceremony in the shattered Executive Mansion, a ceremony held officially to turn the reins over to vice president Moses Blah but really to provide Taylor with a swan song, Taylor delivered a rambling address in which he again compared himself to the persecuted Jesus Christ. Three visiting presidents—Thabo Mbeki of South Africa, John Kufuor of Ghana, and Joachim Chissano of Mozambique—looked on as he spoke, along with a roomful of teary well-wishers.

Later he boarded a plane and flew off to exile in Nigeria. An hour later three U.S. warships appeared on the horizon in a show of force. U.S. helicopters flew over Monrovia. People ran to the beach and cheered.

In late August, after nearly two months of negotiating and politicking, the Accra Comprehensive Peace Agreement was largely worked out. It called for an immediate cease-fire and disengagement among all warring parties; a U.N.-run international stabilization force that included ECOWAS troops; disarmament of all fighters; restructuring of the Liberian Armed Forces and police, and, among other measures, establishment of a Truth and Reconciliation Commission.

Beyond the issue of the TRC—which I will discuss later—the Accra talks had managed to reach agreement essentially by dividing up the government and parceling it out to the various factions. The remnants of Taylor's people received the ministries of Posts and Telecommunications, Health, National Defense, Planning, and Internal Affairs. LURD was given the ministries of Finance, Justice, Labor, Transport, and State. MODEL received the ministries of Agriculture, Commerce, Lands, Mines and Energy, Public Works, and Foreign Affairs. The political parties and civil society groups picked over what remained.

What remained was to choose a new leader, an interim chairman. After much negotiation on just who was eligible to vote, an election was mounted. Eleven candidates, including myself, presented themselves for the position.

There was much backroom politicking during those few days. Deals were made and votes were bought. The Accra corridors were suddenly filled with those who were not part of the dialogue. A great deal of money changed hands—money that came from all quarters but especially from businesspeople who wanted to do business in Liberia and hoped to gain an advantage in government.

In the end the election was held. I received thirty-three votes, the highest number of any candidate. Rudolph Sherman, chairman of the True Whig Party, came in second with eighteen votes, and Gyude Bryant came in third, after a tiebreak vote in which he defeated Togba Nah Tipoteh and Harry Moniba, who had served as vice president under Samuel Doe.

But then the representatives of the warring factions—LURD, MODEL, and Taylor's representatives—met and decided that they and they alone had the right to choose the country's interim leader.

Declaring themselves the "electoral college," they demanded, and received, the names of the three highest vote getters. From these they chose the winner. They chose Gyude.

Rumors swirled that Taylor had sent word that he would accept anyone but me. Whether this was true or not, I'll never know. My belief is that some members of the warring party worried about my reputation for fiscal discipline. This was their chance, as they saw it, to reap the spoils of victory, and they wanted to avoid anyone standing in their way.

Gyude was a political unknown, a quiet, behind-the-scenes team player who had remained in Liberia throughout the worst of the crisis. I thought he would do his best for the nation, and I knew we needed to all pull together at that moment for Liberia's sake. So I accepted the results.

"We have an obligation to the Liberian people to give him all the support we can, and we will," I told every reporter who asked.

On August 18, 2003, the peace agreement was signed in Accra. That October, the National Transitional Government of Liberia (NTGL) was inaugurated under Gyude Bryant. By December the United Nations Mission in Liberia (UNMIL) was fully deployed with a force of more than 15,000 personnel.

The rebuilding of Liberia had begun.

I WAS asked to join the interim government. I had truly wanted and requested the position as head of the Contracts and Monopolies Commission, which would have enabled me to deal with the many entities that had so thoroughly criminalized the economy. Gyude informed me that this position was "beyond your reach." He wanted someone else.

Instead, he appointed me head of the Governance Reform Commission. In the interest of the country, I accepted. It was a critical moment for Liberia, a chance for a rebirth. I wanted to be present to help in whatever way I might.

The Governance Reform Commission was one of several autonomous agencies established under the terms of the Accra peace agreement. Its charge was the promotion and support of good governance in Liberia. Toward that end, the commission was given a broad and somewhat unwieldy mandate, one that ranged from monitoring and reviewing the awarding of national contracts to developing strategies to ensure transparency and accountability in government.

Our job was to be both watchdog of, and parent to, the transitional government, monitoring conditions while helping to create and put into place systems and structural reforms that would move the country away from the abuses of the past and into an era of efficient, transparent, and accountable government. One of my major hopes for the commission was to devise a new structure of government based upon a civil service that was heavily scaled down but better trained and better paid. People who are trained as professionals, taught to consider themselves professionals, and paid like professionals are far less likely to engage in petty corruption.

It was clear from the start that corruption was rampant in the transitional government, just as it had been rampant in every government before. The transitional government was, of course, made up mainly of representatives of the armed factions. Many of those people worked hard and tried their best to make the coalition work, of that there is no doubt. But some saw the time as their opportunity to make as much money as they could. Eliminating this get-what-you-can-get attitude toward public service was, and continues to be, a major challenge for Liberia.

The commission also considered land reform, judicial reform, and the monitoring of government appointments to ensure a certain level of decentralization and deconcentration of power. There was also a public ombudsman role.

We held a series of public interactive forums across the country in an effort to incorporate the views of ordinary Liberians into the reform agenda. And I pressed those ordinary Liberians to advocate for open hearings on the national budget.

Unfortunately, the commission turned out to be a largely policy-making body, with little real enforcement power or authority. Thus when we began raising issues of concern, such as the large amount of government funds being spent on security services for Chairman Bryant, there was little we could do except sound the alarm.

I was, however, able to get one substantive change in the interest of good financial management. We amended the statute to require the constitutional body, the General Auditing Commission, to report to the legislature rather than to the executive.

## CHAPTER 16

# BECOMING PRESIDENT

---

No ONE really expected me to become president.

Among most political observers and operatives, both within Liberia and without, my chances hovered somewhere between slim and none. Conventional wisdom had it that my negatives were too many for me to achieve success: first, I was a woman in a society that insisted on male leadership. I was too closely associated with the settler class. I didn't have enough money. I had been around too long and carried too much political baggage. I was too light-skinned, too educated, and, not insignificantly, too mature to win in a nation so recently and painfully dominated by its youths.

Most media reports pointed to the soccer star George Weah as the almost certain winner. He was young, wealthy, charismatic, and fantastically popular. In a race with such a man I would be fortunate to come in a distant second. At best.

Even most U.S. intelligence and diplomatic reports pointed in that direction. From the time of the Accra peace talks to the final election tally (we learned later), the report from Monrovia to the State Department was that Weah was going to win. According to its calculations, there was just no way that someone as young and popular as he, running in such an environment of young people, could lose. Nor was John Blaney, the U.S. ambassador to Liberia, shy about making it clear where his preferences lay.

"Don't you think this country needs a young, fresh face to lead it now?" he asked me. "So many of you have been involved in politics for a long time. Don't you think you ought to give way?"

"Well, Ambassador," I retorted, "of course the people will elect whomever they want at this point in time. But given the level of destruction and the challenges ahead, I'm not sure we do need a fresh, young, inexperienced face to take over. I'm not sure at all."

The odds then, were solidly against me, but I have always been the type of person who is determined to beat the odds. So in early 2005, I resigned from the Governance Reform Commission to begin my run. Things would be different this time.

For one thing, there was no dominant warlord threatening to send the nation back into bloodshed should he not win. Yes, tensions remained among the factions, and the fear of a tumbling back into war was very much alive. But the cease-fire had held with remarkable success, demilitarization was proceeding, and the UNMIL peacekeepers were working hard to ensure order and stability. Most of all, people were sick of war.

This time too my family was with me in the fight. We all knew— Jennie, Carney, my children and nephews and nieces—that win or lose, this would likely be my last attempt. Everyone also understood that the country stood at a crossroads and that putting the right person into place to lead it forward was crucial. Liberia simply could not afford another misstep.

"We will give you our support," Jennie told me, and they did. My sister, my brother, and many others in my family all adjusted their lives to spend time in Liberia to help me run.

With my decision made and my family on board, my first task was to get the Unity Party organized and ready to work full-time on the campaign. Very quickly I was able to convince Willis Knuckles, a successful businessman and an old friend who had presidential aspirations himself, to put aside his own announced presidential candidacy and come work with me. Willis was known to be a sound manager; when he agreed to join my team, I asked him to head the campaign team.

The next step was to begin mobilizing the financial resources, bearing in mind that our own Constitution and election laws clearly state that non-Liberians cannot contribute to election campaigns. That meant I could not accept contributions from my international contacts. I had to raise the bulk of the money myself and to rely on contributions from Liberians at home and abroad.

In 1997, I had been dramatically outspent by Charles Taylor, who had the immense wealth he had accumulated by pillaging the nation at his command. This time I was determined to marshal what resources I could and to apply them wisely. I did not have a large bank account, nor, as was rumored, a substantial U.N. retirement package against which to borrow. What I did have was property in Liberia. From the time I'd returned to the country I had begun, as had so many, to repair my property. Now I was able to use these homes as a source of rental income and therefore financing. At the same time, several Liberian groups in the United States began to hold fund-raising events on my behalf.

Next I turned to one of my main supporters, Harry Greaves. With Harry I discussed the need for an astute and understanding political strategist, someone to help guide and direct the campaign, to assess the competition and outline a strategy for success.

Harry said, "I have just the man."

Larry Gibson is an American attorney, professor of law at the University of Maryland, and former associate deputy attorney general during the Carter administration. More to the point, he was and is the clear-thinking, exceptionally talented architect of numerous successful political campaigns, including Bill Clinton's 1992 presidential campaign in the state of Maryland and three successful campaigns for Kurt Schmoke, the first black mayor of Baltimore.

Equally important, Larry also knew something about working in African politics. He had run the successful campaign for Marc Ravalomanana, a self-made millionaire who had become president of Madagascar in 2001.

Harry Greaves and Larry Gibson were good friends who had known each other since the 1970s. They kept in touch, seeing one another with their wives frequently. During the peace process of 2003, Harry had mentioned to Larry that he was thinking of returning to

Liberia—perhaps even making his own run at president when the time eventually came. Larry had tried to talk him out of it.

"You have a very comfortable life here," Larry said. "Why in the world would you want to go back?"

"Because it's home," Harry said. "Home is home."

Larry was deeply, and painfully, familiar with Liberia. He had been good friends with Cecil Dennis, the well-liked former foreign minister under Tolbert who had died on the beach with twelve others after the 1980 coup. (Cecil Dennis's widow actually lived with Larry Gibson and his family for several years after that terrible day.) After the death of Cecil Dennis and the deaths of so many other friends and colleagues to the violence that came afterward, Larry had washed his hands not only of Liberia but of much of Africa. He encouraged Harry to do the same.

So when Harry came to Baltimore in early 2005 and told Larry he had decided to support me for president and needed his help, Larry initially refused.

"I have sworn never to set foot in that country again," he said. "I don't want to get involved. Count me out."

Fortunately, Harry is a very persistent man. He went to see Larry's wife, Diana. He gave her what Larry calls the "Obi-Wan" talk, in reference to the scene in the movie *Star Wars* where Princess Leia appears in a hologram and implores Obi-Wan Kenobi to help her because he is her only hope. Harry told the Gibsons that this election was crucial to Liberia and that I was the country's best chance at righting itself and avoiding a return to war, destruction, and death.

"She's the one the country needs," Harry said to both husband and wife. Then, to Larry, "Will you at least meet with the lady and give her some advice?'

That April I flew to the United States to meet with Larry. Traveling alone, I took the train from Washington to Baltimore. Larry met me at the station, and we spent the afternoon talking about what I might need to do to become president. Larry remained highly reluctant to get involved, but I persuaded him to make just one trip to Liberia to conduct an assessment.

"Just come one time," I said. "Go around the country and make an assessment as to whether or not you think I can win. And if I can win, maybe give me suggestions on how to do so."

Larry agreed to spend two weeks in the country, but only on the condition that he would be utterly and completely frank with me about what he found out.

"Am I permitted to tell you if I find out you can't win?" he asked. "Can I be honest with you?"

I told him I expected nothing less.

Larry came to Liberia at the end of May 2005. For two weeks he traveled around the country, talking to people about their hopes and expectations for the next leader of the land. Just days before, the country had completed its voter registration drive and Larry found such a high percentage of registered voters—clearly identifiable by the indelible ink still staining their fingertips—that it surprised him. People were ready, willing, and eager to exercise their democratic rights and to talk openly about what they wanted in a new president.

He interviewed nearly four hundred people: men and women, young and old, educated and not. Among other questions, he asked what each person felt was his or her most important criterion for picking a president: education, experience, platform, party affiliation, or some other thing.

Overwhelmingly, people answered that education was at or near the top of their list. The people of Liberia know that education is the key to the country's future, the means by which their children will raise themselves from poverty. They also understand that the more a person is taught to think for herself, the more she will do so and will not be so easily led. Campaigning in the villages and rural areas, time and again we would hear the people chant, "Who know book? Ellen knows book!"

Larry also found that large numbers of people responded favorably to the idea of a woman president. To be sure, there were some who could not and would not accept that a woman could lead an African nation, such as the man Larry met in Careysburg who just laughed and flat out refused to believe that Larry wasn't trying to trick him by suggesting such a thing. But many more people were not only willing but eager to give a woman a chance.

"Men have failed us," people said over and over again. "Men are too violent, too prone to make war. Women are less corrupt, less likely to be focused on getting fancy cars and fancy home for themselves."

Larry also found that George Weah had very strong support, mostly—but not completely—among the young. He would present serious competition. Still, when Larry came back to see me in Monrovia, his conclusion was firm.

"Oh, yes," he said without hesitation. "You can win."

Larry wrote up a very extensive report outlining the strategy he proposed, what would be needed to put it into place, and the tactics we should take in order to raise my profile above the herd of twenty-two candidates. He did some surveying of the opposition candidates, outlining their strengths and weaknesses and putting forth his thoughts on what shape my campaign should take and the story I should tell.

One of Larry's first—and, I believe, most important—conclusions was that no one was likely to win on the first ballot. The field was far too large and included at least four or five people with significant followings. Besides Weah and myself there were Winston Tubman and Varney Sherman, both attorneys with various fields of expertise. There was also Charles Brumskine, who garnered some early attention from U.S. media outlets such as the *Washington Post*.

Larry felt very strongly that in such a crowded field no one would be able to win 50 percent of the votes on the first try. There was bound to be a runoff, he insisted. This assumption led to several important tactical considerations for us.

First, we did not need to expend all our time, energy, and precious resources trying to win straight out—all we had to do was keep me close enough in the initial balloting to get to a runoff. Second, we needed to be extremely careful about not alienating the other candidates or their supporters. We would likely need them later on.

This was an extremely important point, one that Larry harped on repeatedly and that was sometimes difficult for some of the young and enthusiastic people on my team to keep in mind. Larry believes it was a point that Weah and his followers failed to take into consideration and that it cost him in the end.

For this reason, and also because we felt deeply that the country needed uniting instead of dividing, we decided early on not to run a negative or dirty campaign. We would stay on the issues as much as possible. We would focus on my vision for what Liberia could be. We would emphasize my experience, my international credibility, the contacts I could use to help the nation. We would highlight my record of professional performance and point out that Liberia, still staggering in the wake of such terrible destruction, needed someone who could hit the ground upon election to help the nation heal. Liberia needed a person who could mobilize resources and support very quickly to get the country moving again.

At the end of his two weeks, and after delivering his report to me, Larry said he was packing his bags and returning home. He wished me the best of luck, but that was the end of his commitment.

"Fine," I said. "But before you go, just one more thing."

In Liberia, where so many people live beyond the reach of even satellite television, campaign posters are very important. They put one's image before the voters, make you real in people's minds. I knew the image I projected on the poster would be among the most indelible marks of my campaign. It had to be strong.

Larry, a very talented photographer himself, said he would oversee the photo session and help me choose the right image for my campaign posters. He hired two local photographers and had them picked up and brought to the house one afternoon. They were not told who they were coming to photograph and were instructed that they would be asked to turn over all the film immediately.

Larry had me change into four different outfits—some Western, some African—and strike a series of poses, including—fortuitously— one with my fist raised triumphantly in the air. He had seen an old black-and-white photo of me striking a similar pose before a crowd of people on the afternoon I had come out of prison in 1986. Larry told me later that on the flight home to the United States he had been fiddling around with ideas of the poster when the idea had come to him.

"I still had no intention of coming back, but I did want to offer some suggestions as to what a campaign poster ought to look like. Thinking about that photo of you coming out of jail with your fist in the air, I thought it would be great to design a poster juxtaposing that photo with a recent one, with your hand in the air in the same pose."

That poster became enormously popular. It aligned with the part of our strategy that called for emphasizing my deep experience, bringing attention both to the fact that I had been in government service for many years, going back to the days of the Tolbert administration, and to the fact that, at the same time, I had consistently challenged the government when I thought it was wrong.

Larry also strongly suggested that I appear in my official campaign photo bareheaded, without the traditional Liberian head wrap. He thought it was important for me to signal that I was a modern African woman, connected to the past and tradition but not bound by them.

As Larry said, "African male politicians exercise the freedom to wear either African or Western clothes at their discretion. I thought it was important to signal that you intended to exercise the same prerogatives as the men."

Not everyone on my campaign agreed with this idea. Some were adamantly opposed, but I thought it was a sound idea. Not wearing a head wrap also set me apart from the only other woman candidate. Going out to the villages and talking with people who were illiterate and might be confused by the photos of two women, we could simply say, "This is Ellen. The one without the head wrap. Vote for her."

We flooded the country with those posters (as did the other candidates with theirs). Here again, Larry was tremendously helpful. Because he had contacts all over the world from running so many campaigns, I asked if he would help us secure our campaign materials. He flew to China and arranged for the production and shipping of 850,000 posters, 4 million stickers, 600 banners, untold numbers of T-shirts, and more than 300,000 shopping bags—all bearing some combination of my name and image and the Unity Party logo. In a brilliant move (because it was the rainy season), he also ordered 20,000 rain ponchos emblazoned with the words "The rain will not stop Ellen and the Unity Party." Those ponchos were very popular.

Another early and crucial decision was that I would get out and campaign in every corner of Liberia. I would visit every county at least once, would travel to every city and great meeting place, would show my face in as many villages as possible. In other words, I was going to be the show. This was my own personal commitment. This came from me.

We knew right from the start in which parts of the country we would run strong and in which parts we could scarcely hope to find a vote. I knew I had to win my own home county of Bomi, first and foremost. (Our polling expert, Amara Konneh, had studied the 2000 U.S. presidential election, in which Al Gore had lost his home state of Tennessee and, ultimately, the presidency.)

We knew that in the southeastern part of the country, from which George Weah came, we were not going to be strong. We also knew we were not going to be strong in Grand Gedeh, which was Doe's homeland. There was still, in that area, the lingering idea that I had fought Doe so harshly he could do nothing but fail and that I had set into motion the events that had eventually led to his removal from power. That belief stayed with many in that area, and I knew that getting their support would be an uphill climb.

Still I went there, as well as to the other fourteen counties. I went because we wanted it to be known that we represented all of Liberia and that all of Liberia could count on us if we won. I went because it was important for people to know that we would be fair to every person, every family, every group, and every tribe. We would be the party, and I would be the president, of each and every citizen, irrespective of how he or she cast his vote. That was a message that very much needed to be sent.

As before, getting to some places in rural Liberia was challenging. October is the tail end of the rainy season; there were places where the roads were so rutted or waterlogged as to be impassable, or nearly so. We would drive all night between villages, taking long, painfully slow drives to get where we needed to go. In many places there were no roads at all, and I would just take a canoe and paddle across a river to visit a village. I think that helped my campaign a lot, for people to see that I knew how to handle a canoe, knew about rural life, and was not afraid to get muddy or wet. There were times we drove through mud so deep and sticky we all had to stop and get out and push the vehicle along.

As important as this experience was in winning the election, it was even more important in governing afterward, because I came to know the people of Liberia in an ever-deeper and more compelling way. I trudged through mud in high boots, where roads did not exist or had deteriorated past repair. I surveyed ruined hospitals and collapsed clinics. I held meetings by candlelight, because there was no electricity anywhere—including the capital—except from private generators. I was forced to drink water from creeks and unsanitized wells, which made me vulnerable to the diseases from which so many of our people die daily.

I came face to face with the human devastation of war, which had killed a quarter of a million of our 3 million people and displaced most of the rest. I was forced to hear and understand again—up close and personally—how hundreds of thousands had escaped across borders. How more—who could not—had fled into the bush, constantly running from one militia or another, often surviving by eating rodents and wild plants that made them sick and even killed them.

I was reminded how our precious children died of malaria, parasites, and undernourishment. How our boys, full of potential, were forced to be child soldiers, to kill or be killed. How our girls, capable of being anything they could imagine, were made into sex slaves, gang-raped by men with guns, made mothers while still children themselves.

But listening to the hopes and dreams of our people, I also recalled the words of a Mozambican poet who said, "Our dream has the size of freedom." In campaigning across Liberia, I remembered that my people believe deeply in freedom—and, in their dreams, they reach for the heavens. I knew I wanted to represent those dreams, hopes, and aspirations. In running for president I was reminded time and time again precisely what I was running for.

As hard as I tried, however, I simply could not go everywhere. This is where having the posters and stickers and other campaign materials secured by Larry became so valuable. With them, my staff and large group of supporters, including the amazing market women who mobilized on my behalf, could campaign even in my absence. They went from village to village, meeting place to meeting place, handing out stickers and posters and raising my profile. Larry called this "playing away from the ball."

We also made great use of the Liberian exile community, in which I had much support. Early on, Amara Konneh, the bright young information specialist living in Philadelphia I mentioned earlier, formed a group called Liberians for Ellen and created a Web site to mobilize support for my candidacy.

Konneh told me he'd followed my career since I had refused to take the Senate seat in protest against Doe's stealing of the 1985 elections. In 1990, he had fled Liberia after his father, three brothers, and an uncle were all killed during the war. After some time in a refugee camp in Guinea, he made his way to the United States, where he went to school and earned both his bachelor's and master's degrees. Amara is exactly the kind of bright, determined, and unstoppable young Liberian who will help our country re-create itself.

Konneh became our technical and polling expert. He studied American political operatives such as James Carville and Karl Rove for pointers and learned that successful modern political campaigns constantly take the pulse of voters. In July, before the campaign season officially opened, he flew to Monrovia and conducted a lightning poll of eight hundred young people in the capital to find out how and what they thought of me. What he found was that they knew who I was and even liked me personally. But they were leery of anyone they considered even remotely associated with past governments. Like voters everywhere, they were skittish of politicians because they had been failed by politicians before.

Konneh set up a system that allowed us to track our votes constantly. Using his data, we were able to determine in which parts of the electorate—demographically or geographically—we were weak and why. In August, he conducted extensive polling in six of the fifteen counties, using university students he hired and trained himself. After analyzing the data, he found that Weah had a clear lead, as expected, but that Brumskine was poised to come in second, with me third.

Like Larry before him, Konneh asked voters what they considered the most important issues for the country. Far and away people named education. Presented with this finding, I was able to fine-tune my message to convince voters that I not only understood the pressing need to deliver education to our children but was best prepared of all the candidates to do so.

(Konneh is today at the John F. Kennedy School of Government at Harvard University, completing a master's degree just as I did so many years ago.)

Campaign season got off to a roaring start on Monday, August 15, the first day campaigning was legally allowed. Overnight the capital's streets filled with campaign posters and stickers; overnight every lamp-post, building wall, car, and taxi seemed to sprout an endorsement for this or that candidate. By morning, the excitement was palpable. We held a rolling rally, with six stops scheduled across Monrovia, out to Gardnersville, and back through Bushrod Island. The crowds were so large and enthusiastic that it was hard to move, but we pressed on until I had stopped and spoken everywhere on my itinerary. We drove along in vehicles with sound systems attached, playing our campaign theme song, which was composed by the singer Sundaygar Dearboy. Music is a very important element of Liberian elections.

Speaking to my supporters, I tried to mark both the solemnity and the promise of the occasion.

❝ As a nation, we have endured so many hard times in the past fifteen years. We have suffered. We have lost so much, both in time and in the progress we could have made as a nation and as a people. We have lost family members. We have lost personal treasures. In fact, if each of us were to look back, each of us would have more than good reason to be angry. But looking back keeps us from moving on. We cannot change our nation's past. We cannot move on if we keep looking back. The anguish of war will overwhelm us, and we will do nothing. We must work together to heal our nation's wounds.

The October 11, 2005, elections are the most important elections of our lifetime. In them, you will make a choice about a brighter future for our children and for our country. You will make a choice between hope and fear. You will make a choice between ending the sufferings of our people and worsening their living conditions. You will make a choice between good governance, leadership and transparency, and business as usual. You will make a choice to restore, or not to restore, our country's dignity. **"**

At every stop along the campaign trail I laid out the vision and agenda of the Unity Party. Our vision encompassed a steadfast belief that the current adversity could be turned into an opportunity for national renewal. We believed that peace could be maintained and stability restored, that Liberians could achieve their full potential in an environment that would protect their fundamental rights and guarantee equal opportunity for all.

We believed in a small, competent government that would promote democracy, private enterprise, and the empowerment of people through the exercise of choice and participation. We stressed national unity and promised a government that would reflect the social, religious, and cultural diversity of our nation. We vowed to create an atmosphere free from intimidation and violence and a strong and decisive response to any group that attempted to destabilize the nation and return it to war. We outlined the steps for economic recovery, including job creation, revitalization of the agricultural sector, efficient management of people and resources, and, most of all, empowerment of the people through education and training.

From the start it was clear that we did not have as much money as one or two of the other candidates. In particular, Varney Sherman, the attorney and successful businessman, got his campaign off to a strong and visible start, putting several vehicles bearing his posters onto the streets of Monrovia. Sherman was running as the standard-bearer for the Coalition for the Transformation of Liberia (COTOL), a new coalition of several long-established parties, including the Liberian Action Party (LAP), the Liberian Unification Party (LUP), and the People's Democratic Party of Liberia (PDPL).

Then there was Charles Brumskine, a former president pro tem of the Senate. Brumskine had been a close ally of Charles Taylor but had broken with him in 1999 and fled Liberia. He had returned to seek the presidency and received some early and flattering attention in the U.S. press, especially from the *Washington Post*, which seemed, in some reports, to suggest that his election was essentially a done deal.

Larry predicted—correctly—that Brumskine would be a serious contender, both because of his own personal resources and because of his strong support among many religious groups. The fast-growing Pentecostal churches supported and endorsed him. Brumskine said repeatedly that God had ordained him to be president. He predicted that he would win on the first round.

At one point we decided that we needed to puncture this spreading view. During one joint campaign session at a Baptist church, Brumskine forcefully reiterated his belief that he would be president because God had willed it.

In response I told the audience, "I do believe in God and in his messages. But sometimes we get His timing wrong. In this case, maybe He does want Brumskine to be president—just not in 2006."

All told, our party spent perhaps $2 million on the campaign. Some might say that is too much money to spend on an election in a small country, but everything is relative. And we were far from the top spenders.

The electoral process in Liberia had changed a great deal since 1985. Back then, in many ways, the people themselves actually supported the candidates. You went to a house or a town or a village, and the people there provided accommodations and food for you, maybe even put a bit of money in the pot for travel and gas. By the time we reached 2005, however, the electorate was in no mood to give; they wanted things. In large part this was an outgrowth of the war and the poverty it had engendered; people simply no longer had enough to give.

But they also knew that the competition meant goodies for them, so they expected goodies from everybody: T-shirts, caps, footballs, and bags and bags of rice. One really had to do it. T-shirts were especially popular, because they were clothing and could be worn and worn again. What made it so difficult was that many people became so savvy that it lost real connection to who they were supporting. If they saw the representatives from a certain party coming, they would rush to put on the T-shirt with that party's candidate. If they saw another, they would put on another. They had different T-shirts, and that way they could get as many goodies from as many candidates as possible. For many people, this was a way of scratching a little more at their bare-bones survival and enjoying the largess that was available for such a limited time. I understood this. But it put a financial burden on all of us, because you really had to respond to these things. You could not afford to be left out.

August 15 to October 11 was a seven-week sprint. We went from before dawn until long after dusk every single day, and many times overnight. One thing I have always been is a tireless campaigner. I love it, getting out to talk with people, moving among them, and listening to their thoughts and needs.

Three weeks before election day, I held a major press conference to present my platform and vision for the nation. The event was broadcast live on radio and covered heavily in the press. That was an important opportunity to cut through some of the rhetoric and tell the voters that I understood, on a deep level, the critical issues we were facing, including education, vocational training and rehabilitation for all the young people damaged during the war, improvement of the nation's infrastructure, development of the agricultural sector, and a strong anticorruption element that would bring transparency and accountability to government.

We worked hard to counter the so-called reasons being presented against my candidacy. Some people said the country needed a young person to lead it, not one who might not even be able to finish her term because she was old and sickly and frail! We put the lie to that one by getting me out on the campaign trail, canoeing and walking and dancing with people in the village centers. We also tried to turn it around by saying that chronological age was not the issue. I represented the young in thought and vision, and that was where it counted. What's more, we said, at that critical point in Liberia's history we needed experience more than youthful energy. The country was in dire straits and needed immediate leadership. Now was not the time for learning on the job.

Jennie, in campaigning for me, would put it this way: "Suppose you have a baby," she said. "And you have to go out to work or whatnot and you need someone to care for the child. Would you give that baby to an older person who has taken care of children, who knows how to nurture the child? Or would you give it to a young person who has not done it before and does not know what to do? Well, this country is a child, a sick child, and it is hurting. You need somebody who can nurture it."

It was a simple but powerful message. I think it certainly helped.

Then there were the rumors that I was really a Congo woman, too closely tied to the old settler elite. That one I tried to address by talking about my own history, so that people would know it. I told people straight out that I was of 75 percent rural heritage and the remaining 25 percent was not Americo-Liberian but, in fact, German. More important, the country needed to move beyond the old divisions. We in the Unity Party were there to represent all of the people all of the time.

Finally, there were those who said that a country coming out of war needed a strong person to lead it. A woman, by definition, could not be that strong person, they maintained.

To that I said, "Look at what I have been through. Look and then tell me if there is anything any of these men have been through to suggest they have any better strength of character or purpose than I." The record spoke for itself.

We also pointed out that the country had been led by men for 150 years—and look where that had gotten us.

Then there was one other thing that helped, one other secret weapon up my sleeve. After the 1997 elections, when I had left to settle in Abidjan, I had come back home to Liberia on a visit and gone to Kormah, my ancestral home. It was said that 90 percent of the people in that village—the village where my father and grandfather are buried—had voted for Taylor, but I believed that fear and intimidation had contributed greatly to that number. Taylor's own ancestral home of Arthington is located not far from Kormah, along the same road.

In any event, when I went to Kormah I was taken aback by how devastated the village was, how sad and forlorn it had become. Some ex-combatants had returned home, and the old people were so visibly afraid of these young people it was painful to see. Most farming had been halted during the war; the people were idle, their lives disrupted, their homes destroyed. I knew that something had to be done.

I assembled the residents and talked with them. I encouraged them not to despair or give in, not to sit passively and wait for government or other outside forces to solve their problems, but to recognize that they had the power to do it themselves. All that was needed was the commitment to work together as a community. Thus was born an organization known as Measuagoon, a Gola term meaning "You and I are one" or "We are one."

Working together, the villagers of Kormah decided their first priority was food production. They assigned responsibilities and drew up a list of equipment needed to get started. My resources were meager, but I was able to provide a small fund base of about $1,400 to help the villagers purchase seeds for planting, tools for farming, materials for rebuilding homes, cash to help reopen the schools and to build new wells. That season the villagers cleared and planted some 125 acres of rice, pepper, cabbage, eggplant, tomatoes, and other vegetables. And we expanded the program to include some of the other small and struggling surrounding villages.

When 2005 came around, the people of Kormah and the other

villages remembered the program and how it had encouraged them to retake charge of their lives. Today Measuagoon still exists as an organization. Its reach is small because its resources are still limited, but what we have is always put to strategic use in benefiting the rural people of Liberia.

In mid-September a debate was held at the Centennial Pavilion. Weah and Tubman had declined to attend, which I think hurt them a little. Although there were eleven candidates taking part, most of the questions from the audience were directed to me, Sherman, and Brumskine. I must say I think my showing was best. I took them all on, and my knowledge and comprehension helped gain me support, particularly among the professional class.

All along we kept paying attention to what our polls were telling us. By late September, we saw our numbers rising significantly in the large counties, including Bomi, Nimba, and Margibi. We felt optimistic.

By Election Day we were tired, excited, and nervous. I still believed we would come in second, and then eventually prevail in a run-off, but it was a challenge to be completely confident. Weah's support in the weeks and days leading up to the election was ever visible and always impressive. On the final day of campaigning, his team staged a rally in Monrovia that drew an estimated 100,000 people, most of them young. They marched and chanted and sang, and waved palm fronds and homemade signs, including one that touted Weah as "the Messiah." The crowds were so massive and the day so hot that several people passed out from heat exhaustion and dehydration, and still the crowds pressed on. That was a bad day. For one anxious moment I said to myself, "My goodness. He really is too popular, and our country is too young."

Election day itself dawned hot and humid in most of the country. Massive crowds turned out to vote—men and women, young and elderly people, mothers with babies strapped to their backs. Some people, in their eagerness, slept outside overnight at one of the more than 1,100 polling places. Others woke before dawn and walked miles to cast their ballots.

More than 1.3 million were registered to vote, and officials estimated the turnout at upward of 75 percent. The turnout was so overwhelming the National Elections Commission ordered polling stations to stay open beyond their 6 p.m. closing time to give all the people standing in line the opportunity to vote. Fortunately, we had prepared for this, giving our poll watchers flashlights and sending teams out to fortify them with coffee and biscuits and other food to keep them at their tasks.

After voting in my family's home county, I returned to Monrovia and watched the results begin to trickle in. We were on pins and needles waiting for the vote tallies to come in. Poor Jennie was especially nervous! She said it was like being on a roller coaster ride: one minute strong returns for us came in and we would cheer. Then later strong returns for another candidate would come in and we'd be plunged back down into anxiety. When the bad returns came in people would try to prepare my sister, saying, "Oh, Aunty Jennie, get ready. You won't like to see this!" And Jennie would throw up her hands and go through all kinds of lamenting.

Konneh kept trying to reassure Jennie, and all of us. "Don't worry," he said, because he knew, from his polling, where we were strong and where we were weak. There were no surprises for him. Still, the apprehension was intense. From the start Weah was ahead, I was next and Brumskine was third, just as we had expected.

It would take a week for the results to be complete, though only a few days to get a general sense of where things lay. At one point, Brumskine received a burst of votes and seemed to be catching up. That made us nervous. But then the results from the areas where we knew we were strong came in, and our side maintained its second-place position.

In the end, the NEC announced that, as predicted, Weah and I were the top two vote getters: he with 28 percent and me with 20 percent. Brumskine came in third, Tubman fourth, and Sherman fifth. Since no candidate had achieved a clear majority, there would be a run-off, set for November 8. Campaigning would start on October 27.

Going into the second round there was one thing we knew clearly:

Weah had won, but only by 28 percent. That meant that some 70 percent of the voters were not, at least initially, supportive of him, and did not see him as the person for the job. This was good news for us. We had to go after that 70 percent.

Larry had insisted all along that although Weah's support was wide and passionate, it had little potential for growth. The people who supported him did so fervently, but that group was limited and not likely to expand.

We also believed that the very fervor of Weah's young supporters could end up hurting him. In their youth and inexperience they took on a you're-either-with-us-or-against-us philosophy, which sometimes spilled over into hostility and aggression and ended up alienating the supporters of some of the other candidates. We, on the other hand, had been as vigilant as possible about keeping our team from bad-mouthing or otherwise alienating those on other sides. This required a great deal of discipline, and was not always easy. But because we were firm about doing so, once the runoff was set we could begin immediately reaching out to others.

We knew that one or two of the defeated candidates would swing to Weah, and they certainly did. Although Brumskine—after unsuccessfully contesting his third-place finish—remained neutral in the runoff, most of the others raced to Weah's side. Not only Sherman and Tubman, but also Togba Nah Tipoteh, Baccus Matthews, and Alhaji Kromah all threw their support behind the soccer star.

That was fine with us. We went after the voters, not the candidates. While they were all rushing to Weah's place to pledge allegiance, we re-mobilized our secret weapon—the women of Liberia—and sent them out campaigning. These indomitable women were the real force behind this election, the fuel on which we ran. Singly, in pairs, and in groups they went village to village, door to door, street to street encouraging people to put aside ethnic and tribal allegiances and vote for one Liberia. They held rallies and distributed T-shirts. They walked and danced and marched. More than anything, it was the women of Liberia who turned this election, for me and for themselves.

Anticipating the runoff, we had held back nearly a quarter of our campaign materials. These we used to reflood the country with our message during those frantic two weeks of nonstop campaigning.

One day we left Monrovia early in the morning for a trip to Buchanan, a port city about 100 miles east. We were scheduled to arrive around noon, but in those days of campaigning the schedule was more often a suggestion than a reality. The trip ended up taking the entire day, because at every village and every town and every meeting place along the road the crowds would come out to greet me, often chanting "Ellen! Ellen!," and we would stop.

At every stop we'd be greeted by the village chief or the head schoolmaster or another town official, then I would speak. Just as in the first round, I avoided as much as possible talking about my opponent, instead emphasizing my experience and outlining my vision for a new Liberia, with real focus on the core issues of importance to the people: jobs, education, clean drinking water, reintegration of the former fighting soldiers.

Everywhere we went the crowds turned out. Later, Konneh would tell a reporter, "With my computer program, I could project various scenarios scientifically. But when I could really sense that victory was possible was the day I witnessed people turning out in big numbers in the middle of a rubber plantation at two in the morning."

It was a relentless schedule but I thrived on it. One reporter told Abdoulaye W. Dukulé, a fellow journalist: "I always thought the nickname 'Iron Lady' referred to her indomitable state of mind. I had no idea that it also referred to her physical toughness. She never stops to eat or drink. We have walked tens of miles and driven hundreds more on the worst roads anyone can imagine, but at every stop, she sounds like someone just returning from a vacation."

In the first round some of the other candidates had used a helicopter to get around the country. We only had enough funds to hire one for a few days, so we decided to do so during the runoff campaign. The helicopter enabled me to get to places I would never have been able to reach otherwise. One afternoon we visited all the major cities in Nimba County, which would have taken days by road. We spent another day touring remote Bong County, northeast of the capital.

We also used the helicopter to drop campaign literature over some of the villages in which we could not stop. Later, Larry would say, "What we had not counted on was that some of the people would not see the helicopter or hear it, and that the stickers would get stuck in the trees and drift down over a period of days. It was almost like it was raining Ellen stickers and for some people it was almost like a sign from above. So that worked even better than I thought."

A week before the runoff, Weah declined to take part in a live public debate with me in Monrovia. The debate was to be broadcast on local radio, but his team said he was too busy campaigning to attend. Again, I think not showing up lost him some ground. His absence was read by many as indication that he did not have the capacity to compete face-to-face in a debate.

Around the same time, Weah began suggesting there would be trouble if he failed to win the presidency. Out of nowhere Weah suddenly claimed to have won the first round. He said he had really received 62 percent of the vote, and that it was fraud which had kept him from gaining the presidency. A few days later, the chairman of his CDC party told a gathering of supporters that only cheating by the NEC could keep Weah from winning and that they would not "accept anything less than victory from NEC."

I spoke out forcefully against this kind of talk, which so clearly had the potential of inciting Weah's many young supporters to violence. Others from around the world joined me in doing so, and Mr. Weah did urge his supporters to remain calm whatever the outcome. He did not, however, back down from the assertion that the only way he could lose was fraud. Given the lingering tensions in Liberia from so many years of war, and given the rashness and immaturity of his youthful supporters, this was like laying tinder near an open flame.

On the Sunday before the runoff, I rose, dressed, and went to church. When I returned home, I was surprised and pleased to see that thousands of supporters had rallied, without planning or prompting, in front of my house. We decided to make a last-minute tour of the city, and ended up with a sixty-car convoy that rolled around Monrovia for the rest of the day, stopping in every neighborhood to greet the cheering, chanting crowds. The streets and sidewalks filled with people and together the crowd rolled across the city, ending up at the empty field near my home, some three miles away from central Monrovia. It was well toward dusk by then, but the crowd—happy, excited, expectant—turned into a big rally that went on and on. It was truly exhilarating.

Voter turnout on Tuesday, November 8, was significantly less than during the first round, but the presence of outside observers was equally high. Some 230 international observers from agencies such as the Carter Center joined more than 4,000 Liberians to monitor the voting process.

To keep the peace, U.N. military helicopters patrolled the skies and United Nations Special Representative Alan Doss warned that his 15,000 peacemakers would react swiftly to any trouble. Meanwhile NEC Chairwoman Frances Johnson-Morris called Weah's allegations of cheating "complete foolishness" and warned both him and members of his party against undermining the electoral process.

On Wednesday morning we woke to the news of a commanding lead, with early results from Lofa and Nimba counties, among others, showing me ahead by 60 to 40 percent.

By the end of the week, with some 97 percent of the vote counted, we held a commanding lead of nearly 60 percent. Weah was again alleging fraud, including ballot tampering. Some of his supporters marched through the capital and threw stones at the U.S. Embassy, chanting that the election had been rigged. Later Weah threatened to take his charges to the Supreme Court.

I was never concerned, because I knew the election had been perhaps the freest and fairest in Liberian history. There were so many international observers on the ground—including, but not limited to, representatives from the European Union, the Carter Center, and the National Democratic Institute—that any significant fraud or tampering would have been impossible. There may have been one or two minor infractions, to be sure, but nothing even close to significant enough to alter the vote.

When the NEC called a hearing to examine Weah's claims, neither he nor his supporters could produce any evidence of widespread rigging. Furthermore, representatives from both parties had been present in nearly every polling station and had, on Election Day, signed their satisfaction with procedures. There was simply no indication of the fraud Weah claimed.

As the days passed everyone from the international media to the local religious leaders, everyone from ECOWAS officials to the U.N. representative urged Weah to give up his complaints and accept the results with dignity. As one African diplomat posted in Monrovia told a reporter for allafrica.com, "As an international athlete, he should know more than anyone that one does not win every match. This is his first time and he is still young, but his political future will largely depend on how he conducts himself now."

The truth was that the people of Liberia had spoken. This election truly and rightly reflected the will of the people. It was up to Weah to accept their choice. To his enormous credit, he eventually did.

On November 23, 2005, the National Elections Commission declared me the twenty-third president of Liberia. Weah, again to his credit, attended the official ceremony that afternoon. Receiving the certificate during that moment, I felt a nearly overwhelming sense of pride in, and gratitude for, the thousands of Liberians who had supported me, especially the women.

Most of all, I felt extremely humbled by the faith of all those people and by the task which lay before all of us.

After the ceremony was over, we all went outside and greeted the jubilant crowds. The people whooped and hollered in the streets.

It was a wonderful day.

CHAPTER 17

# Inauguration Day

---

Euphoria.

If there is one word to describe my inauguration day, it would be this: euphoria. I was euphoric. My family was euphoric. My friends and supporters could scarcely contain themselves. More than anything, on that wonderful day it seemed as if the people of Liberia could finally feel and express a sense of hope and joy they had neither felt nor expressed in a very long time. On that day, Monday, January 16, 2006, the very atmosphere of Monrovia seemed to tremble with possibility: the possibility of change and restitution and a brighter tomorrow. On that day I became Africa's first elected woman president. It was a moment seen around the world as one of hope and possibility.

The streets of the capital rumbled with the dancing of the crowds, as people sang and chanted and swayed. There was singing and traditional drumming, and waves of cheers and ululations rippled constantly through the crowds.

Monrovia looked better than it had in years. Thousands packed the grounds of the Capitol Building, which was decked out in red, white, and blue bunting for the inauguration ceremony.

Holding the inauguration at the Capitol Building, rather than at the Centennial Pavilion as was traditional, was especially significant, a way of signaling our commitment to transparency, open government, and participatory democracy for all of our citizens. Nearly two hundred men and women—most of them former combatants—had worked feverishly in the weeks before the inauguration to rehabilitate the war-ravaged grounds as part of a special training program funded by the U.S. Embassy. They had learned carpentry, blacksmithing, masonry and other skills, skills that would help them reintegrate into civil society.

Leaders from all over Africa, and from all over the world, came to share the joy with Liberians and to wish us well. When I stood to give my inaugural speech I could see, among others, Nigerian President Olusegun Obasanjo; South African President Thabo Mbeki; Mamadou Tandja, president of Niger and chairman of ECOWAS; Mrs. Simone Gbagbo, first lady of Côte d'Ivoire; John Kufuor, president of Ghana; Tejan Kabbah, president of Sierra Leone; Blaise Campaoré, president of Burkina Faso; Amadou Toumani Touré, president of Mali; and Faure Gnassingbe, president of Togo. China was represented by Foreign Minister Li Zhaoxing, and Egypt, France, and Finland also sent representatives.

Also in attendance were U.S. First Lady Laura Bush and U.S. Secretary of State Condoleezza Rice. Very importantly, George Weah and several other presidential candidates sat in the front row. Today Condoleezza Rice says the event was one of the most moving of her life.

I took the oath of office, which was administered by Supreme Court Chief Justice Henry Reed Cooper. I swore to uphold and defend our Constitution, to nurture and strengthen our common peace and security, heal our nation's wounds, and lead Liberia to recovery and reconstruction. I promised that with God's help and guidance, I would lead a nation in which we would all put our best efforts and creative energies to work to fulfill our individual and collective expectations.

I meant every word.

The crowd was so jubilant I could barely get through my prepared remarks without being interrupted with shouts of "Yes!" and "Amen!" That first speech as president was very important, a way of setting a tone for what my administration would strive to accomplish. I worked very hard on it. It is reproduced in its entirety as an appendix to this book, but I would like to include here a few of the most important points.

One of the key issues I wished to highlight was that Liberia must always remember and even honor its past while moving beyond it.

❝ THIS OCCASION, held under the beautiful Liberian sunshine, marks a celebration of change—and a dedication to our agenda for a socio-economic and political reordering; indeed, a national renewal. Today, we wholeheartedly embrace this change. We recognize that this change is not just for the sake of change, but a fundamental break with the past, thereby requiring that we take bold and decisive steps to address the problems that for decades have stunted our progress, undermined national unity, and kept old and new cleavages in ferment.                ❞

One of the sections that received the most enthusiastic responses from the audience was this part on the role and future and power of women in Liberia:

❝ UNTIL A few decades ago, Liberian women endured the injustice of being treated as second-class citizens. During the years of our civil war, they bore the brunt of inhumanity and terror. They were conscripted into war, gang raped at will, forced into domestic slavery. Yet, it is the women, notably those who established themselves as the Mano River Women Network for Peace who labored and advocated for peace throughout our region.

It is therefore not surprising that during the period of our elections, Liberian women were galvanized—and demonstrated unmatched passion, enthusiasm, and support for my candidacy. They stood with me; they defended me; they prayed for me. The same can be said for the women throughout Africa. I want to here and now gratefully acknowledge the powerful voice of women of all walks of life whose votes significantly contributed to my victory.

My Administration shall thus endeavor to give Liberian women prominence in all affairs of our country. My Administration shall empower Liberian women in all areas of our national life. We will support and increase the writ of laws that restore their dignities and deal drastically with crimes that dehumanize them. We will enforce without fear or favor the law against rape recently passed by the National Transitional Legislature. We shall encourage families to educate all children, particularly the girl child. We shall also try to provide economic programs that enable Liberian women to assume their proper place in our economic revitalization process.    **"**

After the ceremony, we drove through the city. I got out of the car and walked the streets of our capital, wanting to be among the people. They were singing and clapping and chanting "Ma Ellen!" and other things. The young ones who followed chanted, "Ellen walking; look at Ellen walking." It was sheer exuberance, like witnessing the rebirth of a nation, and I was right there in the middle of it.

I think it was the happiest day possible, not only for myself but for all of us.

On that day I thought back over the course of my life. I thought of the more than thirty-five years I had spent advocating for change. For more than thirty-five years I had worked and fought and risked to make a better Liberia, and finally, here it was: the opportunity to alter—fundamentally—the course of history of my native land.

I thought about my mother and what she would think and what she would say had she lived to see that day. Jennie said, "She would say, 'Well, our baby has finally accomplished this long last goal!' That old man's prediction has come true." Both Jennie and I know that more than anything, our mother would have been praying on that day and every day afterward. She would be praying every single day that arrived that God would lead me to do the right thing by Him, by my people, and by my land.

Because, of course, I knew even on that wonderful day that reality would hit home soon enough. I knew that all those great expectations had to be moderated by the enormity of the challenge at hand.

In truth, the realities were grim. On that glorious day Liberians had not experienced government-provided electric power for more than sixteen years. Unemployment was unacceptably high among all our working populations, but more among our young people who constitute the majority of our population. Pipe-borne water was nonexistent in the city center and parts beyond. Civil servants' salaries were years in arrears and woefully inadequate at any rate. Our schools were in ruins, not just physically but also in terms of the loss of trained teachers. Mortality rates among all sectors of our population were frighteningly high. Our hospitals lacked equipment, drugs, and, most critically, doctors. Our public university system was a shadow of its former self. Our urban roads were terrible and our rural roads all but impassable. Accountability and transparency received only lip-service attention and application in too much of the government sector. Despair and resignation stared many of our citizens in the face. Among some there was little faith that our nation could awaken from its long nightmare or that, in Liberia, government of the people, for the people, and by the people could become a living reality in this generation. All of this was as true on inauguration day as it had been the day before and as it would be the day after.

Later, when I asked Jennie what she remembered feeling on that day, she admitted to a certain ambivalence. "I was so incredibly glad and happy that you had done it, that you had made it, that you'd had the strength to persevere and to triumph in the end," she said. "At the same time, though, I kept thinking 'She's finally made it. Now what? Was it worth all the effort, and what can she do with this humongous task at hand?'"

It was indeed a humongous task before us. I knew that all too well, and so did all of Liberia.

Nonetheless, we celebrated.

On that one day, January 16, 2006, we allowed ourselves the luxury of expectation. On that day there was no restraint for the people of Liberia. On that day I think the whole country felt like "Wow! She will wave a magic wand, and the next day there will be total transformation!"

On that day the sky was the limit. We could achieve anything.

Reality would come later.

# CHAPTER 18

# THE FIRST HUNDRED DAYS

IT WAS a short honeymoon.

Aside from the immediate and nearly overwhelming challenges facing the new administration of a country shredded by sixteen years of war, military rule, nondevelopment, and instability, I was also handed, almost from the start, the very delicate issue of Charles Taylor himself.

Even before the inauguration I began receiving telephone calls from a very persistent U.S. congressman about the issue. Sometime during those short months between the election and the day I took the oath of office, Representative Ed Royce of California called Liberia.

"We're very pleased with this election," the congressman said. "The U.S. supported it. But I want to be very clear with you, Madame President, that if you want your government to succeed you've got to do something about this Charles Taylor. Because we are not going to support this government unless that Taylor issue is dealt with."

That Taylor issue, as the congressman put it, was not, strictly speaking, in my hands at the time. Taylor was living in exile in Nigeria, under the protection of that government, and had been indicted not by Liberian courts but by the U.N.-backed Special Court for Sierra Leone. Still, I knew the issue was out there and that Royce was voicing the sentiments of many in the U.S. Congress and the Bush administration.

"Congressman, we will deal with it," I said to Royce. "But we must have time."

Our first priority was to build a government of inclusion and to begin the very critical process of reconciliation in Liberia. There was still a great deal of disunity, discontent, and factionalization in Liberia. I told the congressman we would have to deal with that first.

"All right," he said. "We'll give you some time. But you better think about what you're going to do."

It was a somewhat unnecessary suggestion, as I did little during those early days but think about what I—what we—needed to do. The people, in a free and fair election, had given my government the greatest opportunity that can come to any leader: the chance to rebuild a nation on the ruins of war. We needed to move resolutely forward toward the achievement of that goal.

But we were far from oblivious to the enormity of the challenges we faced. Few countries in the world had been as decimated as ours. Our human capital was deeply disadvantaged. Half of our people were struggling to exist on less than 50 cents a day. Well over three quarters lived on less than a dollar a day. We had a reported illiteracy rate of nearly 70 percent and an unemployment rate of nearly 80 percent in the formal sector, with a large percentage of those without jobs being young men whose primary skill was waging war.

Our capacity constraints were equally enormous and our infrastructure almost completely destroyed. We took office in a country with no government-provided electricity, no water, and only cellular telephones. Our new cabinet ministers began work in buildings without functioning bathrooms, stripped of furniture and everything else of value. Imagine, if you can, trying to run an efficient, effective, transparent, and accountable government without computers. Several of our senior officials were forced to live in hotels that had a generator and a satellite connection to the Internet, so they could access e-mail and spreadsheets.

Our power grids had been systematically looted of everything, including hundreds of miles of cable. Our water pipes had deteriorated—or had been dug up or stolen. Only a quarter of our population had access to safe drinking water.

In the chaos of war, our HIV rate had quadrupled. Our children were dying of curable diseases such as tuberculosis, dysentery, measles, malaria, parasites, and malnutrition. Our schools—those that remained—lacked books, equipment, and, most critically, teachers. Our clinics and hospitals—those that remained—lacked doctors, nurses, and supplies. The telecommunications age had passed us by.

We had a $4.7 billion international debt owed to the World Bank, the IMF, and other donors, lent in large measure to some of my predecessors who had been known to be irresponsible, unaccountable, unrepresentative, and corrupt. The reality that we had lost our international creditworthiness barred us from further loans—although now we could use them wisely.

Our abundant natural resources had been diverted by criminal conspiracies for private gain. International sanctions, imposed for the best of reasons, still prevented us from exporting our raw materials. Roads had disappeared, and bridges had been bombed or washed away. We knew that trouble could once again breed outside our borders. The physical and spiritual scars of war were deep indeed.

Every segment of Liberian society wanted something from their new government, and they wanted it now.

All the children I met, when I asked what they wanted most, said, "I want to learn." "I want to go to school." "I want an education."

The young men and women I met, our "lost generation," did not consider themselves lost. They, too, aspired to learn and to serve their families and their communities.

Women, my strong constituency, told me they wanted the same chances that men have. They wanted to be literate. They wanted their work recognized. They wanted protection against rape. They wanted clean water that would not sicken and kill their children. These women had not only supported me consistently in my climb to the presidency, but, far more important, they had worked tirelessly during the war to bring the various factions to the negotiating table. The women of Liberia were and remain the country's core, and I have made it a priority to include them in its reconstruction. At the time of this writing, women head the ministries of Commerce, Foreign Affairs, Finance, Youth and Sports, and Gender and Development. Five of the county superintendents are women, out of fifteen.

Former soldiers told me they were tired of war; they did not want to have to fight or run again. They wanted training. They wanted jobs. If they carried guns, they wanted to do so in defense of peace and security, not war and pillage.

The entrepreneurs who were beginning, even then, to return from abroad with all their resources—risking everything to invest in their country's future—told me quite clearly that they wanted a fair and transparent regulatory environment. They wanted honesty and accountability from their government.

Farming families who had fled the fighting for shelter in neighboring countries or found themselves displaced from their communities wanted a fresh start. They wanted to return home. They wanted seeds. They wanted farm implements. They wanted roads to get their goods to market.

As I had come to power peacefully, I had to keep the peace. As I had campaigned against corruption, I had to lead a government that curbed corruption. As I had been elected with the massive vote of women, I had to ensure that their needs were met.

I had many promises to keep.

Despite all these pressing needs, our first priority had to be to organize a government of inclusion and then to get started on developing an agenda for the six years of our term. The country had been so badly divided during the conflict that there was simply no way to even begin addressing the country's many challenges without first bringing us together in some way. The election process itself—with so many different candidates appealing to so many different constituencies in the first round—had exacerbated the fragmentation that had so undermined us in the past.

It was very clear to me, based on my own experience as an opposition leader in 1997, that unless you can draw people to you, either they are going to go into oblivion, thus wasting their contributions and skills, or they will become an active force working against the agenda you try to set. Sometimes opposition is good for the sake of democracy, but in a time of recovery it can be detrimental if the opposition seeks to undermine what one is trying to accomplish. Especially if these are people who have a certain amount of credibility or popularity in their own right; in that case, their unhappiness can be quite a distracting force.

I knew it was crucial to begin to reestablish a sense of ourselves as one people and one nation, all working together for the common good. Liberians first, everything else second.

So one of the first things I did upon taking office was to ask George Weah to be part of the government. He declined but said he'd like to make sure I put some of his people into positions of authority.

We tried to do just that. Unfortunately, it was very difficult, because most of his people did not have the proper qualifications. During those first hundred days, as we were putting together our government, we came under some criticism for not including more people from this party or that. But I was clear from the beginning that even though I wanted very much to create a government of national unity, there was a firm criterion for doing so: all those who came to work with us had to have competence in their field; they had to have commitment; and they had to be clear of any record of established corruption.

Using that criterion, I was able to reach across the spectrum and take into the government several presidential and vice presidential candidates. That was an important step. Not only did those people bring their constituency into the coalition we were trying to build, but by including other candidates we were able to send the signal that despite the fact that they had opposed me, as long as they had the ability to be part of the team, they would be included.

Of course, the job of trying to promote reconciliation and unity is only just begun. It will take many years to create a truly unified Liberia. But during those first few months we were very busy both trying to begin this process and to develop the agenda that would carry us through six short years.

At the same time, we launched the Truth and Reconciliation

Commission of Liberia. The nine-member commission, which was designated under the 2003 peace agreement, was charged with examining human rights abuses from January 1979 to October 2003.

In announcing the formation of the commission, I said, "We must make collective restitution to those victimized, rehabilitate the victimizers, while at the same time visiting some form of retribution upon those whose violations qualify as crimes against humanity.

"This commission is our hope—to define the past on our behalf in terms that are seen and believed to be fair and balanced, and bring forth a unifying narrative on which our nation's rebuilding and renewal processes can be more securely anchored."

The TRC set about to gather testimony, primarily from victims but also, ideally, from low-level perpetrators, including the children forced to join the fighting and kill, not only in Liberia but among exiles in Ghana and the United States as well. There are those who say, understandably, that more is required than simply attribution and forgiveness, that justice itself must be applied. Those people want a war crimes tribunal. They say that those who committed heinous crimes must be brought to justice in a court of law and face whatever punishment is prescribed.

There was much discussion of this point during the peace talks in Accra, when the mandate for a TRC was first established. For obvious reasons, many of those participating in the peace talks were far more strongly in favor of a TRC than a court that would seek criminal prosecution for the commitment of wartime atrocities.

Still, the decision to move forward with a TRC went beyond our just giving in to the clear intent of the warring factions to avoid prosecution for the things they'd done. During those long years of conflict in Liberia, the madness of war engulfed so much of the country and so many of the people that we knew if we decided to send everyone who had engaged in such actions to court we'd be holding trials for two or three decades. Thousands upon thousands of young people, many of them children as young as ten years old, were forced into violence. They were given alcohol, which their young bodies could not manage, let alone their young minds. They were drugged and then, in their drugged and bewildered states, forced to do horrendous, incomprehensible things. In Accra we knew all this, knew it painfully and firsthand. We asked one another, "What do we do with these young people? Do we take them, thousands upon thousands, to a war crimes tribunal? Do we hold them to a level of accountability they could not possibly have been held to when the nightmare began?"

We all agreed that, as a first step, we needed to put into place a process of contrition and forgiveness. We decided that a war crimes tribunal would serve little purpose, would be very expensive, and would consume a lot of time and resources so desperately needed to rebuild our shattered land. The TRC seemed to be a better alternative: we needed to have all those young people who caused so much pain stand before their accusers and face them and apologize and then go on and try to make new lives for themselves. When their accusers say they want more than forgiveness, they want justice, the TRC must recommend justice for them. Then we can talk about who should face a court. It will be, then, a two-step process. But our first step was to find the means to rehabilitate those who were conscripted into war. It is not a question of undermining justice. We were, and are, trying to find a balance between justice on the one hand and reconciliation on the other.

We knew it would not be easy, and that has proven to be the case. The TRC has run into some very rugged moments, with the hearings sometimes threatening to become more accusatory than confessional and with those who then stand accused rising in denial, emotions high. But we are trying to move forward in the process. We have really just begun.

I also spent some time during those first few months traveling the world as an ambassador for Liberia, rallying the significant international support the country would need to rebuild itself. In March, I traveled to the United States to address Congress—only the second Liberian president to do so after William Tolbert—and to meet with President George W. Bush in the Oval Office.

That first official visit to the White House was exhilarating, a major moment for both me and Liberia. President Bush had actually sent word shortly after the election that he would like to meet me but protocol demanded that the meeting wait until after I was inaugurated. The president came across as a genuinely down-to-earth person, warm and likable, as is his wife, Laura. I also met with Condoleezza Rice and others. We had very good talks about the future of Liberia.

Even with all of this and more on our plate, we found that the Taylor issue kept threatening to crowd everything out. Not long after the inauguration I paid a visit to Nigeria to personally thank President Obasanjo for all that Nigeria in general, and he personally, had done for Liberia. During our meeting I brought up Taylor and the pressure we were receiving from all quarters to resolve the situation.

President Obasanjo said, "Madame President, you know my position. It is quite clear."

I did. Even before the election President Obasanjo had declared, in response to pressure from the United States, that Taylor would be turned over to the United Nations only if there was evidence that he had violated the asylum agreement. Absent such evidence, Taylor would be deported from Nigeria only if the head of a duly elected government asked for him to be so. It was a stance that both bought the president some vital time and tossed the hot potato of Charles Taylor into someone else's hands.

"Nothing should be done to erode the credibility of Nigeria," President Obasanjo told the *Washington Post* in May 2005. He explained that he had consulted widely with other nations before granting political asylum to Taylor. If he reneged on the asylum agreement, Obasanjo said, "nobody will respect us."

During our meeting, President Obasanjo told me how Taylor had visited him more than once since his arrival in Nigeria. I think this personal relationship made it difficult for the president, because, like so many of the other African leaders, there was for him a certain amount of attachment to Charles Taylor. He liked him. So although he knew full well that Taylor was bad for Liberia and bad for the region as a whole, part of him wondered if this man could possibly be reformed. I think many of the African leaders could not keep from asking themselves, even at the end, "What can we do to save him? Can his charisma and his strength be used for good?"

Of course, Taylor's own actions quickly dispelled any lingering hopes of his redemption. On the one hand, talk floated of how even more deeply religious he had become, going to church every day and donating huge sums of money to the church. At the same time his former associates, some banned from travel by the U.N. Security Council, began boldly trekking back and forth between Liberia and his home in exile. Openly they bragged about having been to see "Pape" recently and about going back again and again. That Taylor was trying to reorganize his people was very clear.

During our meeting, President Obasanjo said that during one visit Taylor had gone so far as to suggest that all President Obasanjo had to do was put him in a plane, fly him quietly and secretly to the Liberian bush, and put him down. "I will take it from there," the president quoted Taylor as saying. "Leave it with me. I will take it from there."

We both chuckled at that suggestion.

During the election, some of Taylor's people from the NDPL had supported my candidacy. Among the most prominent, of course, was his wife (or former wife; their exact relationship remained unclear), Jewel Taylor, who, after her election to the Senate in October had endorsed my candidacy during the runoff.

I received some serious criticism for accepting her support. Some people asked, how can you possibly work with her? My position, though, was that here was a woman who needed to establish her own identity. Jewel Taylor said she had divorced Charles Taylor, and I took her at her word. I thought she needed to walk away from her husband's shadow and become a person and a leader in her own right. So she supported my candidacy, and I accepted her support.

Later the relationship turned sour, because she had felt the support she gave me would be in exchange for saving her husband from being turned over to the Special Court. We did not have a deal like that. In her own mind she thought we had that kind of a deal, but I never suggested or agreed to such a thing.

Nonetheless, because some of his people had supported my candidacy, they seemed to believe it gave them carte blanche. They began to get a bit bold about communicating with Taylor in exile and relaying his words back to his young supporters at home. His favorite statement, the one he had uttered before leaving Monrovia and the one that rang over and over again in the months that followed, was this: "God willing, I will return."

Meanwhile, the president of Sierra Leone was understandably anxious for the situation to be resolved. And the pressure from the international community continued to mount. Whenever I met with U.S. officials, the issue of Taylor was on the agenda. Likewise with representatives of the United Nations. By that time the Special Court had been in session for several months. A great deal of money was being spent to run it. People wanted to know if events were moving forward or not.

In March 2006, I went again to Nigeria to meet with President Obasanjo.

"Mr. President," I said, "this time I am going to ask you officially to release Charles Taylor."

He said, "Okay. You must send me a letter to that effect."

In that letter, and in a press conference making our request public, I was as clear as I could possibly be about this one point: Nigeria was not releasing Taylor to Liberia. The Liberian government did not have a case against Taylor. Nigeria was releasing Charles Taylor to the international community, as represented by the Special Court for Sierra Leone.

This was an exceedingly important point, because Liberia's peace was still very fragile and Taylor still had many loyalists on the ground. It was frustrating to me that the decision about what to do with a regional problem such as Taylor—who had gone into exile under an agreement negotiated by several West African leaders—had been put squarely on the back of Liberia, a small, damaged, and still fragile country. To have my administration involved in his jailing in any way was risky. I asked President Obasanjo to rally the other African leaders to make a joint formal statement requesting that Taylor be tried. I knew that once the request was made public, time would be of the essence in securing Taylor. All in all, it was a risky move.

Finally President Obasanjo announced that Nigeria would hand Taylor over to Liberia. I asked that Taylor be sent directly to Sierra Leone, but Obasanjo was insistent that he would return Taylor only to Liberian soil. In releasing his statement, the president said he had consulted with ECOWAS leaders as well as with President Denis Sassou-Nguesso of the Republic of the Congo, then head of the African Union.

I do not know all the details of how Taylor was taken into custody. Suddenly the newspapers were reporting that he had disappeared. Whether this was true or part of the plan to distract Taylor and his supporters during the capture, I cannot say.

What I do know is that on the morning of March 29 we received a phone call simply saying "Taylor is on his way." We quickly, and quietly, mobilized the U.N. peacekeepers and got them to Roberts Field International Airport. We deliberately did not use any of the African troops but asked for the Irish troops instead. The troops were put on high alert, and the United Nations had a helicopter standing ready. Our plan was to have Taylor land on Liberian soil and be taken immediately into U.N. custody.

Somehow some people in the media got wind of what was happening and raced to Roberts Field International Airport. However, as Heaven would have it, just as the plane carrying Taylor landed, a heavy downpour took place. The rain was so relentless it kept the reporters from swarming the plane. The U.N. troops used the cover to move Taylor quickly from the plane and onto a waiting helicopter. I'm told that Taylor asked to go to the restroom but the U.N. commander said no. He knew you couldn't trust Taylor not to take advantage of the slightest opportunity.

By the time the downpour eased and word was out that Taylor had returned to Liberia, he was over our airspace and on his way. We telephoned the president of Sierra Leone so that he too could prepare. When Taylor reached Freetown, he was immediately jailed.

To say that I and many others breathed a sigh of relief upon being told that Taylor had left our airspace would be the understatement of the year. Had he escaped while in the country, there would have been big trouble for everyone.

In an address to the nation following Taylor's transfer from Liberia, I stressed that his fate was now in the hands of the international court and that the United Nations should and would allow him to maintain his dignity and the right to a vigorous self-defense, consistent with the principle that a person is deemed innocent until proven guilty. I also stressed—because it was important to do so—that no other Liberian had been indicted by the Special Court for Sierra Leone, and no one needed to fear for his safety or freedom.

Furthermore, I said:

❝ I ALSO want to tell all of those who accompanied or lived in Nigeria with Mr. Taylor that, as citizens of Liberia against whom there is no charge, they are free to return home and will be given the same protection and opportunities as any other citizen so long as they remain law abiding.

Nevertheless, the Government wishes to make it abundantly clear that those who try to use these circumstances as an excuse for insurrection to undermine the stability of the nation will be dealt with harshly and without mercy, as prescribed by law. While the Government will respect their rights as provided for in Articles 15 and 17 of the Constitution, they should be mindful of the provision of Article 76 of the Constitution.

Fellow Citizens, as you know this Government inherited the matter relating to Mr. Taylor and try as we did to avoid it, in the end I had no option but to accept the responsibility of leadership, by taking the hard decision which ensures the long term safety of the Liberian people and the security of the state. This is what I pledged to do when I took the oath of office on January 16.

Today, we can just thank God that with his blessings, this saga for Liberia has come to an end and the government can now pursue what it was elected to do—bring development and prosperity to the Liberian people. ❞

Taylor was safely imprisoned in Freetown. But we all knew he could not remain there long. There would be too much traffic back and forth, too much opportunity for bribing guards or sending messages or plotting escape. Moreover, his very presence continued to undermine stability. Sierra Leone was as vulnerable as we.

Even before Taylor was taken into custody, I had been making the case, during meetings with the U.N. Security Council, that Taylor must be taken from African soil and imprisoned at The Hague. This was the only way to guarantee the fragile peace and security of the entire subregion.

As soon as Taylor reached Freetown we began working on plans for transferring him to The Hague. But progress stalled because the Dutch wanted a guarantee that Taylor would not be jailed there if convicted. They wanted a third country to step up and offer to take him. No one volunteered. Austria was approached, as were Denmark and Sweden—but no dice. Some members of the United Nations began asking why the United States and Britain did not offer to jail him themselves.

Finally, after months and months of diplomatic maneuvering, the British did just that. They offered to take Taylor should he be found guilty and to provide a place for him to serve whatever sentence he might receive. With that promise in hand, the Dutch agreed to the transfer. In June 2006, Charles Taylor was flown from Africa to the Netherlands and jailed in The Hague.

Even in finally leaving the continent, Charles Taylor left fresh damage in his wake. My relationship with President Obasanjo suffered some strain after he came under fire from other African leaders for turning Taylor over to the international community. Some criticized him privately for taking what they considered an African problem to be solved by Africans and turning it over to the United Nations. Libya's Muammar al-Gaddafi went so far as to publicly lash Obasanjo for "betraying African solidarity."

Yet I also believe that many leaders understood that if Africa is going to truly build democracies it must learn to recognize when one of its own has failed, and to take action when that happens. Even as the Taylor crisis was coming to a boil, southern Africa was dealing with the ongoing problem of Zimbabwe's president, Robert Mugabe. The Zimbabwean government has been severely criticized by the international community for corruption, economic mismanagement, and violation of human rights. Yet the leaders of Africa know that Mugabe is far too strong to be challenged. If an example needed to be made—and it did—that example would have to be Taylor. There was, beneath the surface, a very strong push to make an example of one African leader who had violated his own country's laws and infringed upon his own people's rights.

In the end, Taylor had become a symbol of evil. People across the continent were deeply angry with him not only because of what he had done in Liberia, but even more because of what had happened to the poor people of Sierra Leone.

The brutality during the conflict in that nation was simply staggering—the raping and breaking of family ties, the hacking off of limbs. Whenever photographs of the lingering effects of that war were shown in the media, it raised people's desire not for revenge but for justice to be served. People felt that Taylor simply had to pay the price for such wanton death and destruction. Every time a newspaper ran a photo of some young man or woman marred for life by the loss of his or her limbs, it increased the continent's resolve: we cannot let this one go.

In the end, though, Taylor was a temporary distraction for our new and fledgling government. There were many predictions of trouble, many predictions of doom as we struggled to resolve this lingering part of our troubled past. People said, "Oh, if Taylor gets taken from Nigeria to Freetown, there will be rebellion! Chaos will erupt! People will march in the streets!"

Thankfully, that did not happen. Thanks be to God, that was not the case.

We did what we did because it was the right thing to do—and we felt the majority of the Liberian people wanted nothing more than to put the Taylor era firmly behind them. They weren't only tired of it, they were, as our African-American brothers and sisters say, sick and tired of being sick and tired of it. We only asked and continue to ask that he be given a fair trial with the right to self-defense and that his dignity as a former head of state be respected. The court is doing just that.

With the issue of Taylor behind us, we could concentrate on our development agenda. We established a list of steps toward poverty reduction to be taken within 150 days, using the framework of an Interim Poverty Reduction Strategy, a staff monitoring program supported by the IMF and World Bank. The program has four pillars: Peace and Security, Economic Revitalization, Infrastructure and Basic Services, Governance and the Rule of Law.

These important sectors are where our attention was and remains focused. Although we knew Taylor had a fervent group of loyalists who would make a lot of noise and try to stir up tensions, who would whisper and shout and get on the radio raising a ruckus, we also believed that ruckus would be short-lived.

The majority of the people, we thought, would simply dust their hands of the past that Taylor represented and look forward. The majority of people, we hoped, would be more than ready to move on.

Which proved to be precisely the case.

# CHAPTER 19

# SOME CHALLENGES AHEAD

DURING MY campaign I often said that one reason I was running for president was because I wanted to see the children of Liberia smile again. They have gone through such horror: running from guns, facing starvation, seeing their parents mutilated and killed. During my campaign, as I traveled the country, I could still clearly see the signs of despair, the loss of hope. Too many children still did not know where their next meal was coming from. Too many still knew nothing about school.

So my greatest commitment during this time as president is to change that situation for our country's children, to fulfill the promises I made. Not only to make them smile; that's a bit superficial, of course. My real commitment is to get them decent food, decent housing, decent education. To get them back into an environment in which they can feel themselves to be a real and vital part of a real and vital society. To create for the children of Liberia an environment in which they feel the future is bright.

For this to happen, our children must go to school. As we strive for national reconstruction and renewal, education of all the children of Liberia, especially the neglected girl children, must occupy a place second to none in our national priorities. Our democracy simply cannot be advanced when the majority of our citizenry is functionally illiterate and lacks the knowledge and skills required to lead our national efforts and to compete in a global community.

We are working to enforce the country's Free and Compulsory

Education legislation, which has been on the books since 2004 but never fully funded or implemented, for many reasons. Among them is the simple and frustrating fact that there are not enough schools in Liberia to educate all of our children—not enough schools, not enough teachers, not enough resources.

If you ask a former child soldier—and there are thousands and thousands in our country—"What do you want?" the answer in every single case is "I want to go to school." Schools need to be staffed now. These scarred children are our responsibility and our future. We must move them from victims into a state where they can grasp the future.

In our first year under the Free and Compulsory Education program all tuition and fees were abolished in public primary schools and tuition and fees in public high schools were reduced and standardized. In my first year of office we created a 40 percent increase in school enrollment. We put emphasis on the girls, because traditionally girls have been neglected—not only in Liberia but in all of Africa. When parents have limited resources with which to educate their children, they tend to put the emphasis on their sons rather than their daughters. Traditionally this made sense: boys were the ones who went to work, girls got married and became mothers. Obviously, times have changed and we will too.

One means of redress we have created is the Liberian Education Trust, a program that is mobilizing resources from individuals, corporations, and institutions abroad. Our aim is to build or rehabilitate 50 schools, train 500 teachers, and give 5,000 scholarships to young girls. As of this writing, we have raised some $3 million and begun construction on nearly twenty schools.

We also passed new regulations making it a punishable fine for children to be on the street doing petty trading during school hours. Our goal is not to punish parents but to encourage them to view their children's education as a priority, even at the cost of what are undoubtedly needed extra funds.

HELPING OUR children is one thing. Helping our young people, many of them ex-combatants, is another issue entirely.

As of this writing, through our demobilization and reintegration program, we have processed some 120,000 former combatants, but there still remain some 30,000 or so to be pushed through. (This is in addition to the 17,000 persons demobilized under the deactivation and restructuring of the Armed Forces of Liberia, the Liberian National Police, and the Special Security Service. This action, begun under the National Transition Government of Liberia, cost an estimated $20 million. We have now begun reorganizing those institutions and re-cruiting and training new personnel.)

The hard truth is that a good number of these young former combatants—our children, yes—are hardened criminals. They have known violence for the better part of their lives, and combating that truth is very difficult. Theft and even armed robbery remain serious problems in Liberia, and these hardened former combatants are largely the source of it.

Sometimes a foreign investor will complain to me of theft from the site of some building or hotel he is constructing, and I will sympathize and try to address the situation, but I will also say to him: you must understand, for decades the people of Liberia lived in a society of de-pendency. After that they endured a period of political turmoil and the introduction of violence into everything they did, as a result of which our people became accustomed to living in survival mode.

The human instinct to survive is powerful; when operating at that level, people will do whatever it takes to make it from one day to the next. This is still the mentality that drives the actions of too many of our young people. This is still what we are dealing with. Too many of our young men, especially, have not learned to think through the im-plications of what they do, not only for their own futures but for the process of peace and nation building for all of Liberia.

We are struggling to move our citizens beyond this short-term men-tality. Our challenge is to make them understand and believe they have a future, and that whatever they do today must be done in the context of tomorrow. It will not happen overnight. We must give them both the means and the opportunity to pull themselves away from this way of thinking and also create punishments for continuing to operate in that mode. We cannot do one without the other and hope to succeed.

One solution is not only to complete the reintegration training of as many former child soldiers as we can and as quickly as possible, but, most important, to create jobs. So many of our infrastructure programs, such as road building and housing construction, we try to make as labor intensive as possible in order to create jobs for these young people. More important, we are working diligently to rebuild the economy, to get the mines started again, and to get the forestry operations started again and the rubber farms up and producing as quickly as possible.

These industries, once in full operation, will absorb many of those young people, both pulling them out of the desperately overcrowded and overburdened Monrovia and giving them a sense of purpose and accomplishment. We have to believe that once these young men get good jobs where they are earning a decent living in exchange for their hard work, crime will diminish. We must believe that most people want only the means and the opportunity to become good, productive citizens and that, given that much, they will do the rest.

This, then, remains one of our biggest challenges: to find the approaches, policies, and mechanisms whereby citizens can play a much more important role in the economy. That is the real means of generating sustainability in an economy: not just giving people jobs but being able to support enough growth and entrepreneurship to get them to move from petty trading and medium-sized businesses to a place in which they can create enterprises large enough to compete on the international stage, enterprises that can obtain and make use of the best technology and capital. That's a big challenge, especially in Liberia, where the economy has, for so very long, been based on large, extractive industries owned and managed by foreign nationals and dependent on a large commercial class, again managed by foreigners, mostly Lebanese and Indians.

Where do you even begin in such an economy? Where do you begin when you have so many Liberian businesspeople and entrepreneurs, the very people with the potential to build your economy, no longer even living in the country, having fled because of the conflict and pain? Once the conflict has ended, how do you rebuild? At the same time, you want to create an environment that subscribes to international competitiveness, fairness, and equity. You cannot be discriminatory, because to do so will cut off the very capital and technology that is needed to rebuild the economy and to grow it. Liberia must find the right balance between opening its doors to foreign investment and creating a safe and fertile space in which to begin developing its own indigenous entrepreneurs.

To say our first years in office have been challenging would be an understatement. One of the most difficult challenges—one of the toughest things in Liberian culture in general—is simply creating the capacity to get things done. In the more developed Western countries, once you identify a problem and find the financial resources to address it, the problem is all but solved. But when you are dealing with a population, many of whom have been bypassed by education for so long, who have been deskilled through years of war and inactivity, when you are suffering from a brain drain because most of the talent has long since fled the country, just finding the people and the wherewithal to accomplish all that you must is an enormous task. There you stand, trying to rebuild a nation in an environment of raised expectations and short patience, because everyone wants to see change take place right away. After all, they voted for you precisely because they have confidence in your ability to deliver—immediately. Only you cannot. Not because of the lack of financial resources but simply because the capacity to implement whatever change you have in mind does not exist. You can't blame it on inefficiency, and you can't blame it on laziness—the people simply do not possess the necessary skills. This is one of our greatest challenges: developing the capacity of our own people to do all the jobs a functioning democracy requires. You can have short-term on-the-job training courses, and we are doing that; you can send lots of young people away for formal education abroad, and we are doing that too. But all this takes time. And time is what we are short of in Liberia. Time is a luxury we do not possess.

In these efforts, and in others, we hope to be aided by various international groups. Among the partnerships I would most hope to develop would be one between the people of Liberia and African Americans.

The example that Liberians use is of Israel and the Jewish community in the United States. In the Jewish community in the United States Israel has a strong and effective constituency that regularly supports and advocates for Israel. Liberians look toward this example and say to themselves, "So. If there is any country in Africa that has a historical connection with a group of Americans, it is us and African Americans. By their very name this group has declared that its roots are African. And ours are theirs."

Yet this natural constituency is not there for Liberia—or, rather, it has not been in the past. More than once, a white American in the U.S. Congress has told us, "Your problem in Liberia is you don't have a constituency in America. There is no one here working on your behalf."

One hundred years ago, or seventy-five or even fifty, one might easily have understood why this was. African Americans were still struggling and fighting for their own rights as citizens in America, for equal access and opportunity. But now, as a whole, they have reached a point of being competitive in their society. They have broken down the barriers and now enjoy the same kinds of privileges as any other Americans. Our white friends in Congress tell us bluntly, "It is they who should be reaching out to you. If there is anybody who should invest in Liberia, it is African Americans who should take that leading role."

We understand, however, that a historical tension remains between African Americans and Africans. We as a people became slaves in another country because of our own. We as a people did not volunteer to go to America and become slaves—we were sold into slavery by other Africans.

Moreover, I believe there is a segment of black Americans who want to distance themselves from the settlers of Liberia, because of what the settlers did in forming this land. They left a land of bondage and oppression and came here to impose another form of colonialism on the people already here. Who wants to be associated with that? The whole stigma of slavery and oppression that Liberia represents—that Africa represents—to many black Americans is precisely what they want to distance themselves from.

When African Americans do turn to Africa to claim it, they tend to want to attach to places other than Liberia—places such as Ghana or Nigeria or South Africa, places that do not carry the stigma of a settler class creating a society so divided it led first to a coup and then to war. They don't want to deal with that. It's much easier to go to Ghana and say, "I come from here." It's much easier to go to South Africa, with its long and noble history of fighting oppression by a white minority, and say, "I'm from the Zulu tribe." To their credit—and let us not forget—when we were fighting apartheid in South Africa, it was the strong African-American community that put a lot of pressure on the United States to help end that oppressive system.

(Even Oprah Winfrey reportedly thought she was Zulu until the eminent African-American scholar Henry Louis Gates, Jr., informed her that she is most likely descended on her maternal side from the Kpelle people—of Liberia!)

The natural bond that exists between Israel and Jews, then, is not exactly the same for us. It will take a great deal of effort on both sides to get across that historical distancing. But things are changing, and bridges are being built. People are saying "Hey, we're not going to change our history. But that history does not have to determine what we do together from here on out."

During my first year in office I met with the African-American entrepreneur Robert Johnson, who told me of his plan to assemble a delegation of prominent African Americans from the worlds of business, politics, and entertainment who would all join in an initiative, part of the Clinton Global Initiative, to invest at least $30 million in Liberia. He brought along the Overseas Private Investment Corporation, a U.S. agency which helps U.S. businesses invest overseas, and also invested $3 million of his own funds. Needless to say, I was delighted and welcomed him aboard the train. This wonderful initiative is, I think, part of a significant and growing constituency for Africa among African Americans.

In April 2007, Johnson led a twenty-five-member delegation to Liberia. The group, which included the actress Cicely Tyson, visited businesses in Monrovia, toured villages in the countryside, and met with Liberians from all walks of life. A year later, I broke ground for the construction of a new, $10 million, four-star hotel outside Monrovia being built by Johnson. Construction of the hotel will create between 300 and 400 much-needed jobs in the area and help attract other development and industry.

Said Johnson after his 2007 visit, "We were awed by the challenges but moved by the sense of hope and faith Liberians have in their future. Every Liberian with whom we spoke said that the country will not return to war. Liberians want to rebuild their lives by finding jobs, restoring their homes, and educating their children."

He is right, of course. This restoration of hope is one of the best things about being president of Liberia, one of my greatest satisfactions. Just yesterday I was coming out of somewhere, a program of some kind, and as I left the building I saw the children lined up to greet me. As I almost always do, I had my driver stop the car so I could get out and greet the children, hand out some biscuits and candy, touch their hands. There was one little girl who simply could not stop herself from jumping up and down, so excited was she. For her this was such a joyful day in her life that I stopped the car and got out and I went to them and had them sing the national anthem. That touched me; such moments are visible, beautiful signs of a restoration of hope to Liberia.

I am also encouraged by the signs of progress we see all around us. Every day I see changes taking place: the restoring of basic services, the rebuilding of infrastructure. I just feel so good about that.

Also, I take heart from the new Liberian image abroad. Before, the only thing people seemed to know about Liberia was that it was a failed state, a place where people killed one another, a land where death and destruction characterized most of life. During those long, dark years,

Liberians were really ashamed even to present their passports at various immigration posts, for fear people would look at their passports and wonder: Who are these people? Who am I letting into my country? Are they dishonest people trying to find refuge because they have pillaged and destroyed their own home? Before, Liberians traveling abroad found themselves under all kinds of suspicion. But today Liberians tell me they are very proud to show their passports. People take one look and say, "Oh, yes! How is your country doing? How is your president? We saw her on the news, and we wish you well!"

Today Liberians are proud to be Liberians again.

## ON AFRICAN VALUES

In Africa the debate between Western individualism and African collectivism rages on. Some call for a return, in Africa, to the concept of *ubuntu*—a Bantu word that expresses the concept of the traditional African value of collectivism, of emphasizing the communal good over individual desires and needs. Others say the belief in a golden era of African collectivism is exaggerated and that clinging to such economic concepts in the modern world is self-defeating.

I believe that in the old Africa collectivism could and did work. In that Africa the individual economic spirit was not as strong as it is today; capitalism, and whatever it means in terms of the accumulation of individual wealth, was not part of the value system.

But as populations across the continent have increased, the pressures on land have become more intense. Land is a big issue everywhere in Africa and a major one in Liberia. We are moving with all due speed toward the establishment of a land reform commission, to deal with squatters and the very troubling problem of illegal sales. My office has received so many complaints from landowners it is astonishing. People who have a little piece of land sell it to three or four different persons with the same mother deed, then sit back with the money while the purchasers go to court to fight it out. This is unacceptable.

There is also the issue of our traditional system of collective land-ownership in the rural areas, a tradition that goes back to the tribal system. That has proven damaging to agricultural production, because when nobody owns the land, nobody takes responsibility for working it. We are trying to move toward a system in which that land is distributed to people in the village and they hold fee simple title to it. This will enable them to use it as they want, either to put to productive use themselves or to sell it. But land reform is a major challenge for our administration. Moving some of the old chief and tribal leaders from a system in which they as chiefs had complete say over who got what piece of land into a more modern arrangement is not going to be easy. To say the least.

Returning to the idea of collectivism: globalization and the radicalization of societies have led more people to embrace the idea of getting ahead on the basis of their own industry. This is the Africa of today, and there is no going back. Governments must respond to changing times and come up with new measures to increase economic prosperity without creating new tensions. In truth, governments must sometimes force old traditions to give way to new requirements.

That said, there are without doubt elements of the traditional African value system I would not want to see disappear. First and foremost is the idea of the extended family. That means, among other things, having concern for the aged of your family, taking them in, and giving them haven. I would hate to see things here such as nursing homes, where parents are left to fend for themselves. So far, thank goodness, we still maintain that tradition: we take care of our mothers and fathers.

All around Monrovia today there are high walls and barbed wire, lots of barbed wire. Someone is making a killing from selling barbed wire, and it is a depressing sight to see. We are afraid of one another now. Whatever the problems of yesteryear—and they were many—that was never true before.

I do not think we'll ever recapture that spirit of total ease among one another. I do not think we'll ever go back to a time when people could feel free to walk the streets at night or stop in on a neighbor's yard unannounced. We've probably lost that forever. But to the extent we can inculcate into people the idea that you have a responsibility to your neighbor, that you are, in fact, your brother's keeper to a certain extent, that will be good for us.

A GREAT challenge for the entire world is the continuing tragedy of war. Our own recent past makes us acutely aware of the suffering of civilian populations—still an acute agony for Africa. In eastern Congo, where 4 million people have died in the past decade, the largest contingent of U.N. peacekeepers in the world is working to curb the violence. They deserve our attention and our support.

And it is past time—as I said to the U.N. General Assembly in 2006 and yet, as of this writing, must say again—to stop the killing in Darfur. There is no excuse to delay action because of disagreement over whether the instrument should be an African Union or U.N. force. The continued stalemate exposes weaknesses in international cooperation and collaboration and a lack of international will to address the suffering.

Civilized nations must not be indifferent to any conflict—internal or external—regardless of the factors that fuel it. We must properly equip and fund the current African Union mission now, while we move urgently to mount an effective U.N. intervention. Our government has called upon the General Assembly and the Security Council to exercise its authority, under Chapter 7, to restore peace, security, and stability to Darfur.

As an international community, we must also work harder to curb the tools of war. Liberia suffered under economic sanctions, because wood from our forests and diamonds from our lands were used to fund conflict. We have now passed a new forestry law and joined the Kimberley Process Certification Scheme, a U.N.-recognized certification system for the sale of rough diamonds. We have understood our responsibility to police diamond exports and believe that diamond-importing countries must better ensure that they are not customers for conflict diamonds.

Similarly, we must work together more effectively to limit the trafficking of weapons to illicit users. Manufacturers, exporters, and purchasers of small arms must join forces against those who kill and maim and rape, who create the instability that is hospitable ground for terrorism. My administration has taken a visible role in the fight to end the proliferation of small arms and light weapons. In November 2007, Liberia took part in a conference of thirty countries from West, Central, and North Africa held in Tunis, Tunisia, to discuss small-arms control, and we continue to push the issue. Liberia supports the International Action Network on Small Arms in its campaign to gather support for an Arms Trade Treaty, a legal instrument that would prohibit arms from being exported to destinations where they are likely to be used to commit grave human rights violations. Such a treaty—which is essential for nations all over the world and especially so for those of us in Africa—would require countries to comply with international human rights and humanitarian law standards when authorizing weapons transfers.

Furthermore, if we truly seek a peaceful and just world, we must systemically enhance women's access to—and participation in—decision-making processes. We're proud to say we have made a beginning in Liberia, with women leading key ministries and agencies, including our national police and the ministries of Foreign Affairs, Finance, Commerce, Youth and Sports, and Gender and Development. But while women assume their rightful positions of leadership, both domestically and internationally, we must not forget that the poorest and least educated people everywhere are women and children. I have long supported the establishment of a new, independent U.N. fund or program for the empowerment of women and for gender equality. Again, as I stated at the U.N. General Assembly in 2006, the United Nations must demonstrate its commitment to the world's women by establishing such an institution with adequate funding and a mandate to carry out this responsibility.

## On Building Democracies

Rebuilding democracy, especially in a postconflict environment, is a painstaking and arduous process. Today's mature economies, such as the United States and many European countries, can themselves attest to this. Reconstruction has to be inclusive, all-embracing in its character. In countries such as ours with three distinct branches of government, each branch is acutely aware of its domain and sphere of influence. Yet engagement at all times is a sine qua non. And there can therefore be only one sure path to that end—old-fashioned consultation, constant dialogue, and information sharing and dissemination, at all times. And when compounded by the imperative of postconflict national reconciliation, the task is great for all actors, all our stakeholders—government, civil society, including our embryonic but energetic media, the private sector, and of course our external partners.

Second, in seeking to move away from our emergency, humanitarian phase to a more developmental phase, our people will have to make fundamental shifts in their own attitudes—in living, in doing business, in coexisting with one another. They will have to break with time-honored traditions if, for instance, these are violations of human rights or perpetuate systemic inefficiencies. They will have to become less dependent on the state, as the state may no longer be able to afford it. Breaking away from these is never easy and often politically sensitive. We will also need to understand that public service means just that—serving our people and doing it efficiently, honestly, and openly. Postconflict development will require new and imaginative and at times more sympathetic ways of doing business. A key to this monumental shift in attitudes will be our continued commitment to massive investment in the education of our people, to which we remain steadfast.

Third, we will need to accept that good-neighborliness and good neighborhoods are mutually reinforcing realities, with serious inherent challenges. We in Liberia are part and parcel of a resource-rich subregional neighborhood with strong traditions of cross-border movement and commerce. We have to sustain those in mutual respect while exploiting the many synergies of political infrastructure and other economic and social cooperation. As history has time and again clearly manifested, an unstable Liberia is an unstable Mano River subregion (defined as the area around the Mano River and including Liberia,

Sierra Leone, and Guinea). And an unstable Mano River region is an unstable West Africa.

Fourth, and very important, we will have to seriously confront a number of challenges that attend the rebuilding of strong strategic partners, domestic and especially external. For us, good partnerships are two-way streets. Mutual accountability by both government and partners is a sine qua non. For their part, our partners must minimize responses of support driven by ideology or geopolitics. Rather, their responses must be largely tailored to the complex postconflict realities and architecture on the ground. Our partners must be creative and responsive in their mechanisms for identifying, with us, our necessary support and mutually agree upon delivering that support in a timely way.

As it currently stands, more often than not, by the time the international community arrives, it is usually nearly too late. In Liberia, for example, most international observers arrived a month or so before the election, set up their observation teams, witnessed the election, announced their findings, and then left. This was helpful as far as it went—but it was not nearly far enough.

What is needed is more forward thinking, preventive action, and longer terms of commitment. The international community must commit to a far more sustained effort to support and consolidate democracy in underdeveloped countries, particularly when those countries are in a postconflict situation. Because the record is clear: unless countries receive that kind of sustained support, support that is both substantial and timely, they will not succeed. Whatever disunity caused the conflict in the first place is still there; the fracture remains, and it must be addressed at its core.

Our international partners must quickly, and in relevant ways, build up capacities on the ground, adequately empowering their staff to make on-site decisions as much as possible. This way our postconflict countries can quickly have access to the necessary material, technical, and financial support, thus assisting us in eschewing what the literature is today calling the "conflict trap"; our social partners themselves often entrepreneurs, the NGOs must be accountable for their work and be prepared to be more open and transparent; our private sector more accommodative, supportive, innovative, and responsible to the realities of postconflict environments.

The root cause of conflict is not simply poverty but poverty brought on by exclusion. Exclusion in its broadest context: exclusion from resources, from power, from education and information, from the opportunity to better one's life. All of these exclusions contribute to poverty. And unless we have programs that address it and do so quickly, whatever euphoria and raised expectations come from a good election can all too quickly dissipate.

Then you're back to peacekeeping, which is incredibly expensive. What is the American saying? An ounce of prevention is worth a pound of cure. At the time of this writing, the international community, led by the United Nations, is spending nearly $1 million a day for peacekeeping operations in Liberia. Just think if that money were put into development!

That, then, is one of the messages we've been trying to send during these first years of our government. Liberia has been lucky in this regard. We have received this kind of support from the international community. In some cases it has not been as timely as we would like, but because we have a wide range of support from private individuals and corporations, the latter has filled the gap. In truth, our chance of being different, of being able to consolidate this democracy, has come about largely because of my own extensive contacts. I have been able to call upon those individuals, and they have responded positively to try to help us succeed.

But of course the world can't count on each struggling democracy electing an individual who has those kinds of connections. We need a more systematic response, a more structured program for building and supporting democracy in developing nations. The only body that can possibly lead such an effort is the United Nations.

Even to say this, however, runs the risk of playing into certain perceptions of Africa. There is the long-standing perception that Africa will not grow up, will not accept its own responsibility. In this view, the people and nations of Africa are always looking for help, always depending on others, always expecting to be bailed out. In this view, Africans have failed to manage their own resources properly, mostly out of laziness or incompetence; in the cases where they had the opportunity to do so they were too interested in aggrandizement and the accumulation of personal wealth. Meanwhile, anyone with the capability and means to do so fled to seek his fortunes in a place with more opportunity. This, then, is the reason Africa has failed to develop and evolve and will continue to fail. There's nothing much you can do with this continent. Better to wash your hands of the place.

Of course, the unfairness of all of this is that Africa's evolution was interrupted by two enormously destructive events: slavery and colonialism. We cannot use those historical realities as an excuse for some of our laxity, shortcoming, and failures, but nor can we pretend that they were not both enormously important elements in who and what Africa is today. Slavery robbed Africa of millions of its people—the young, the healthy, the fit of spirit, body, and mind. How else would they possibly have survived the horrors of being marched to the coast, carried across the Middle Passage in the dark, disease-ridden holds of slaving ships, and then placed into bondage? Slavery severely limited the continent's capacity, and colonialism came afterward to further the damage. The scramble for Africa, and the Berlin Conference of 1884–1885 that formalized it, carved up the continent for European exploitation. King Leopold II of Belgium took the Congo and ran it as his personal farm! The Berlin Conference and the colonial scramble arbitrarily divided some ethnic groups, forced others together, and created a legacy of political fragmentation that still reverberates. Colonialism also put into place a system of authoritarianism that favored elitism. Even after both systems had been dismantled, their legacies remained.

In the first, heady years of postcolonialism, Africa essentially followed the same colonial pathways that already existed, simply replacing colonial leaders with revolutionary ones. Then the continent suffered another interruption in its evolution through the spasm of military takeovers, because the revolutionary leaders failed. (It is important to note that once again the West—fighting its Cold War and still seeking to exploit Africa's resources—fueled some of the problems with our first revolutionary leaders.)

Also missing was education. As recently as a generation ago, some countries had barely four or five Ph.D.s in the entire country with maybe a few master's degrees to spare. Most of the people were illiterate, and even the bulk of those in power had stopped their education in high school. Without the human capacity to properly understand and manage a nation's resources, corruption began to play a more problematic role.

It has taken almost two generations after the end of colonialism for Africa to realize its problems and come to grips with them. Likewise, it has taken two generations for Africans to begin to realize the enormous potential the continent holds. Only now is Africa beginning to take charge of its own destiny. Only now is the ship slowly turning around.

The path ahead remains long, and the technical know-how and capital needed to move Africa to the next level are still, at this point, dominated by the West. Whether those resources are provided by the IMF or the World Bank, they are still controlled by Western powers. And there is a feeling among some in Africa that the West does not really want the continent to become truly independent. A strong, united, independent, and productive Africa would mean a challenge to the prevailing economic structures of the world. It would mean increased competition in the marketplace for those currently providing the bulk of goods and services.

In fact, there is a growing fear about China for just this reason. China is the coming tiger, and it has clearly identified Africa as an area of major strategic and economic interest. The Chinese influence in Africa can be seen in the capitals of many countries any day of the week.

What are the implications of this? Does that put Africa once again into rivalry between the superpowers? There is no collective African response to these questions as of yet. We are all thinking the issue through.

Whatever happens with China, we in Africa realize we cannot break our dependency on the West in as short a time frame as we would like. Some countries, such as South Africa, in which the economy was dominated and controlled by a powerful white elite and in which technology, technological skills, and education were transferred to the general population, have viable, working economies.

The rest of Africa is still struggling along, moving slowly forward, inch by inch. We have not gotten the big boost that the devastated European nations were given by the Marshall Plan, a boost that moved them from ravaged lands to institutional power. We probably will never be given such a boost.

Nevertheless, the challenges faced by Liberia and much of Africa are by no means insurmountable. But in confronting them we need to draw strength from ourselves and our rich past as a people, as well as from the time-honored successes and continued goodwill of our enduring partners. I am confident that Liberia can and will become a nation to be proud of once again.

And Africa will rise.

# CHAPTER 20

# THE FUTURE

---

I GUESS the debate is still on over whether leaders are born or made.

I'm frankly not sure how one develops the skills necessary to lead and lead well. Leadership requires stamina. It requires a whole lot of acceptance, the ability to remain committed to your cause and to have the courage of your convictions. It requires understanding that sacrifices will have to be made—and the willingness to make them again and again and again.

The greatest sacrifice of all is putting everything important—the challenges, the needs, your own ideals and sense of responsibility—ahead of yourself. In effect, to be a great leader is to sacrifice oneself, because if you ever stop to think about your own preservation, your own safety, and your own survival, you will immediately become constrained. You will cease to act, or to act in the best interests of those you are leading. To be a great leader means to get to a place where personal considerations and needs become secondary to the achievement of your goal. That is the greatest sacrifice you can make, but that is precisely what leadership demands.

It seems reasonable to think that certain people are born with these qualities of leadership but that their experiences as they go through their life's journey serve greatly to add to and strengthen them.

Every aspect of my life, every event, every challenge, every threat helped to mold me into the person I am today. My strong and faithful mother, my charming and ambitious father crippled in the prime of his life, my marriage, my motherhood, my education and professional experiences—all these people and experiences molded my character in some way, shape, or form. Even going to prison was, I believe, a great lesson in leadership. Going to prison in Liberia did not make me angry or bitter; it made me incredibly remorseful and sad. It made me feel that way because at the time I had already held several high positions in government, yet I had no idea about the conditions that existed in my country's prisons, about the ways and methods in which ordinary people ended up there. I was a leader of people, yet I did not know what those people were really going through. None of us did—didn't know or didn't care or didn't make ourselves care. We never even thought about what was going on in the prisons of Liberia, never wondered: Who was there and why? How long would they be held? What would it take for them to get out? What kind of conditions were they in?

Sitting in prison day after day, coming to the understanding that if I were going to help the people, I needed to know, really know, their lives, was a humbling and important experience for me. Whenever I was inclined to feel sorry for myself or afraid I would reflect, "Hey! Poor people go through this all the time!"

Perhaps this should be part of the proper grooming of leaders: to be put into a position where you suffer what the common person suffers. How else can you really understand what you're working to do?

I understand. I see this government as the true transitional one for Liberia, the one that will introduce the reforms necessary to rebuild society—both superficial and structural. I feel I have a mandate to do just that.

The deep-rooted structural problems we have faced and are facing— socially, politically, economically—are things we desperately need to address during these short six years. Roads and schools are not the things that really create unity in a society. What creates unity is a common identity, a common sense of purpose and understanding.

Unlike other African countries, which had a common cause around which they could rally—defeating apartheid, ending colonialism— Liberia has never had a real sense of national identity. We have never, in our history, been sure whether we were true Africans or just transplanted Americans. Even though the majority of our population are in fact of indigenous, native descent, over the years the imposed culture of superiority of the settler class caused much pain and made too many long to be like them, to be somehow both African and not.

In the later years, a reversal came, of course: more and more indigenous youths became educated and radicalized. This was the basis of the coup, but once down that path we still never attacked the basic problem of what it means to be Liberian. Who are we? What are we? What are our true roots and what is it that we should be working toward? We still lack a spirit of nationalism that will represent a renaissance. It is that renaissance that we all yearn for today but just can't bring ourselves to fully embrace, because we're still so individualistic in everything we do—clannish, tribalistic, divided among ourselves. We've never risen above that to see the national good as that which binds us and is superior to anything else we are or represent. Until we can do that, we will always be threatened with falling back into disunity.

Today, thank God, we have peace and have had it for five years, yet the society is still torn apart. We are still so suspicious of one another's motives and agendas. We don't trust one another because we think each person is only out to snatch an advantage for himself, his family, his tribe. We are suspicious of everything everybody else does. It's very sad, and it is the most challenging part of trying to unify this nation and really put it on the path of sustained peace and development.

I wish I could say I have found the means of ending this suspicion and bringing us all together, but I have not.

Some of our citizens call for a national conference to address these problems. But the truth is, we've had national conferences in the past. People come, they present nice papers and make eloquent statements, and then we all go back into our little cocoons with our same old attitudes. Before we hold another national conference, we must first prepare for it with real conversation, real debate on the subject. We need to talk openly and honestly all over the country about our history and our roots, about the evils of our past and the good in it too. We cannot change history—we have to accept it for what it is and stop the finger-pointing and the blame. And we must move ahead into the future together.

We have begun this process through a program called "Changing Minds, Changing Attitudes," which we hope will be the beginning of trying to focus on some of the issues. The Truth and Reconciliation Commission is another important part of the process. All of these efforts must continue, and must succeed, for the sake of our children, so that they do not grow up in the same stratified society that has characterized us for so long.

At the same time, however, I know I cannot focus so much on these large, philosophical issues that I lose focus on the smaller, more superficial issues facing the people of Liberia in their everyday lives. The superficial is fixing the roads and the power and the schools and the industries that create jobs. These issues are visible, immediate, and quantifiable, and to ignore them is to introduce disappointment.

I have to find the right balance between addressing our deep-rooted cleavage but at the same time responding to the everyday needs, which is important to meeting some of the expectations of the people. And I have to do it in six years—not a long time.

In six years I will not be able to carry the reform to the place where we are on an irreversible track. There will be a lot of progress yet to be made. That is why I consider this a transitional government.

I know I will not finish the critical tasks at hand in this term—and therein lies another challenge. How can I ensure that the next government, with or without me, will stay on track and continue the reform? This is not an academic question. If we were to get into a bad government again, which is all too much a part of our history, we would have a lot of reversal. And that reversal could set us right back on the track to disunity and, God forbid, conflict.

What I have to do is ensure that I meet the needs of the people as far as possible, so that they see a better future ahead; launch serious and fundamental reform; and then determine how I can best influence the next government so that those reforms become consolidated and complete.

And after that? Well, perhaps I will earn the Mo Ibrahim Prize for Achievement in African Leadership, a new prize rewarded annually to the retired African head of state who most significantly improves the lives of his or her country's citizens. The winner, who must have also democratically transferred power to his or her successor, receives $5 million over ten years, then $200,000 annually for life thereafter. What is more, the winner may also receive $200,000 a year for ten years toward her public-interest activities and good causes. Former Mozambican president Joachim Chissano was awarded the first Ibrahim Prize in 2007, but there's no reason a woman cannot win soon.

I am increasingly looking forward to that chance.

IN KATHARINE Graham's memoir of her life and times as publisher of the *Washington Post*, she wrote that she was often asked why she never remarried after her husband died. It seems to be a question that is commonly asked of women who succeed in the professional or political world. Certainly it has often been asked of me.

My answer is that I married when I was seventeen years old, and I took my husband's name. By the time we were divorced, I had begun, with much hard work and effort, to establish a professional name for myself. Everything I had achieved, everything I had worked for and accomplished, had been accomplished and achieved in the name I took when I was seventeen. To change it would have constituted a kind of starting over, and that I was unwilling to do.

Also, my children bore the name I had taken. I wanted to retain that part of my life with my children, that very important connection to them. Had I remarried, especially in those early years after the divorce, when my children were still young and at home, it would have introduced another element into the relationship between mother and child. Even in the best of circumstances, stepfamilies can be challenging to navigate.

Then too, during those early years I was very much in pursuit of professional excellence; I didn't want to get bogged down again in a marriage that would lead to a conflict between my professional and personal lives. That is not to say that I did not have romances, because I did; I had one very special relationship. Having a relationship without marrying enabled me to have closeness to and connection with another person, but in the context of my own choosing and at my own time and pace. It allowed me to pursue my professional goals without having to make the sacrifices necessary for a marriage. It allowed my children to maintain their primary place in my life.

No mother can fully ever state the importance of her children in her life. Because I had my children when I was very young, in some ways I was more like a big sister to them than a mother, especially during the early days. In a way, we grew up together, my boys and I; we played ball together, we swam, rode bikes. But because the family also broke up and because I was separated from them so often as I tried to build my career, there was, and still is, some disconnect with each of them, except perhaps for Rob. The truth is, they have also borne a lot: the worry and the upheavals of my career. When I was thrown into jail, they had no means of support; they were in school and essentially had to fend for themselves, to get out and see how they could raise the funds to continue their schooling. The youngest was at Morehouse at the time, trying to get his doctorate. He had to take out student loans that he is still trying to repay.

But I thank God that all my sons came out of those times very well. Given the kinds of pressures they had to face, from struggling for money to even being jailed on my behalf, it would have been easy for any one of them to have slipped into drinking or drugs or gone bad in some other way. Instead, they have all become professionals in their careers and wonderful men in their personal lives, and today we are, all of us, very, very close. Even Adamah, who for many years, I believe, felt I should not have left him when he was only one year old—I think even Adamah has forgiven me now. I am proud and honored to be an icon for my sons, to be the model who survived their father and moved up the ladder of success. To be the foundation for their own success.

I have tried, honestly, to keep them out of politics, though I suspect that one or two of them, most likely my son Charles, will eventually go that way. He and my first son came home to Liberia to help rebuild the country, and they are working hard to do so, but the financial rewards are not what they might be in the States. So I help them as much as I can, supporting some of the grandchildren who are now in college. Adamah, who is a physician, remains in the United States but visits often to bring physicians and experts to help improve our health care system, while my son Rob also remains in the United States but is indispensable to me in many ways. We are all very close in that regard, and I am grateful. I am happy I've been able to give back to them in these latter years what I was not able to give them in their early years, when I was going through my political and professional trials.

So I have no regrets about not remarrying. In fact, I can honestly say I have few regrets at all when I think back over my life. There is nothing I would have done differently. Everything I did, every action I took, matched the circumstances of the times. If I could live my life again, I would live it exactly the same way.

One other question people sometimes like to ask is whether I believe I would have accomplished more or less as a man. I don't have to hesitate to answer that one—I would have accomplished far, far less. I would have been, really, just another man. I think that as a woman I was an exception, and being an exception gave me both the visibility and the drive to succeed. I was ahead of my time, but I am no longer alone. We are breaking barriers daily; in another decade there will be hundreds of women in real positions of leadership all over Africa and all over the world.

I take pride in having helped trample down those barricades. I have been one of the lucky ones.

# Inaugural Speech
## by Ellen Johnson Sirleaf

---

LET US first praise Almighty God, the Arbiter of all affairs of humankind whose omnipotent Hand guides and steers our nation. Before I begin this address, which signifies the high noon of this historic occasion, I ask that we bow our heads for a moment of silent prayer in memory of the thousands of our compatriots who have died as a result of years of conflict.

Thank you! I also ask your indulgence as I reflect on the memory of my two rural illiterate grandmothers and my mother and father who taught me to be what I am today, and the families who took them in and gave them the opportunity of a better life.

Let us also remember in prayers during his affliction, His Grace Archbishop Michael K. Francis, the conscience of our nation. Vice President Joseph N. Boakai and I have just participated in the time-honored constitutional ritual of oath taking as we embark on our responsibilities to lead this Republic.

This ritual is symbolically and politically significant and substantive. It reflects the enduring character of a democratic tradition of the peaceful and orderly transfer of political power and authority. It also affirms the culmination of a commitment to our nation's collective search for a purposeful and responsive national leadership.

We applaud the resilience of our people who, weighed down and dehumanized by poverty and rendered immobile by the shackles of fourteen years of civil war, courageously went to the polls, not once but twice, to vote and to elect Vice President Joseph Boakai and me to serve them.

We express to you, our people, our deep sense of appreciation and gratitude for the opportunity to serve you and our common Republic. We pledge to live up to your expectations of creating a government that is attentive and responsive to your needs, your concerns, and the development and progress of our country. We know that your vote was a vote for change; a vote for peace, security, and stability; a vote for individual and national prosperity; a vote for healing and leadership. We have heard you loudly, and we humbly accept your vote of confidence and your mandate.

This occasion, held under the beautiful Liberian sunshine, marks a celebration of change—and a dedication to our agenda for a socio-economic and political reordering; indeed, a national renewal. Today, we wholeheartedly embrace this change. We recognize that this change is not just for the sake of change, but a fundamental break with the past, thereby requiring that we take bold and decisive steps to address the problems that for decades have stunted our progress, undermined national unity, and kept old and new cleavages in ferment.

As we embrace this new commitment to change, it is befitting that for the first time in our country's 158-year history, the inauguration is being held on the Capitol Grounds, one of the three seats of Government. We pledge anew our commitment to transparency, open government, and participatory democracy for all of our citizens. Yet, we are humbled and awed by the enormity of the challenges that lie ahead—to heal our nation's wounds, redefine and strengthen its purpose, make democracy a living and effective experiment, promote economic growth, create jobs, revitalize our health and educational facilities and services, and quicken the pace of social progress and individual prosperity in this country.

My Fellow Liberians:

Today, as I speak to you, I wish to state that I am most gratified by the caliber of the delegations of Foreign Governments and our international and local partners who have come to join us to celebrate this triumph of democracy in our country. I am particularly touched by the presence of the African Union Women Parliamentarians and others of my sisters, who are participating here with us today in solidarity.

I wish to pay special recognition to several African Presidents who are here today. His Excellency Mamadou Tandja, President of the Republic of Niger; His Excellency Chief Olusegun Obasanjo, President of the Federal Republic of Nigeria; His Excellency John Kufuor, President of the Republic of Ghana; His Excellency Thabo Mbeki, President of the Republic of South Africa; His Excellency Tejan Kabbah, President of the Republic of Sierra Leone; His Excellency Blaise Compaoré, President of the Republic of Burkina Faso; His Excellency Amadou Toumani Touré, President of the Republic of Mali; and His Excellency Faure Gnassingbé, President of the Republic of Togo.

All of you, especially the Leaders of the Economic Community of West African States (ECOWAS), have spent invaluable time, energy, and the resources of your respective countries to help guide and support the process of restoring peace, security, and stability to Liberia. To General AbduSalam Abubakar and his team, we thank you. We adore and respect you for your persistence and commitment in the successful implementation of the Comprehensive Peace Agreement which gives closure to 14 years of civil conflict with my taking the Oath of Office today.

My dear Brothers and Sisters of West Africa:

You have died for us; you have given refuge to thousands of our citizens; you have denied yourselves by utilizing your scarce resources to assist us; you have agonized for us; and you have prayed for us. We thank you, and may God bless you for your support to Liberia as well as for your continuing commitment to promote peace, security, stability, and bilateral cooperation within our sub-region—and beyond.

·Permit me to take special note of the presence of Her Excellency Mrs. Laura Bush, wife of the President of the United States of America, Her Excellency Condoleezza Rice, Secretary of State of the United States, and other members of the American delegation. Mrs. Bush and

I share a common passion and commitment to gender equity and the education of the girl child. I salute her for her work in Africa and in the Persian Gulf Region. I also thank her and Secretary of State Rice for their presence—and support. For us, this manifests a renewal and strengthening of the long-standing historic special relations which bind our two countries and peoples.

It also reflects a new partnership with the United States based on shared values. We are confident that we can continue to count on the assistance of the United States and on our other development partners in the urgent task of rebuilding of our nation. We note with satisfaction the presence of Ms. Louise Frechette, the Deputy Secretary-General of the United Nations; His Excellency Cellou Diallo, Prime Minister of the sisterly Republic of Guinea; His Excellency Li Zhaoxing, Foreign Minister of the People's Republic of China; His Excellency Hans Dahlgren, Special Representative of the European Union to the Mano River Union;

His Excellency Louis Michel, Commissioner of the European Union for Development and Humanitarian Aid; His Excellency Alan Doss, Special Representative of the Secretary-General of the United Nations in Liberia; His Excellency Dr. Ali AbduSalam Tiki, Special Representative of the President of Libya and Minister for the African Union; and all other distinguished delegates to this inaugural ceremony.

In acknowledging your presence, permit me to express through you to your respective Governments our deep appreciation for your moral and financial support and contribution that have enhanced the process of restoring peace, security, and stability to Liberia. I wish to acknowledge the stewardship of the National Transitional Government under the leadership of its former Chairman, Mr. Gyude Bryant, for their contribution to peace and to the successful electoral process. I also recognize and thank the former National Transitional Legislative Assembly for their service to the nation.

And I welcome the members of the 52nd Legislature who were sworn in a few moments ago, and are here resolved in Joint Assembly. Distinguished Ladies and Gentlemen, I congratulate you as you assume your individual responsibilities of representing our people. I look forward to working with each of you as we strive to build a better nation.

I thank and applaud our gallant men and women of the Armed Forces of Liberia who have rendered sacrificial service to our nation and are now being willingly retired to facilitate the training and restructuring of the new Armed Forces of Liberia. I also thank the leadership and gallant men and women of the United Nations Mission in Liberia (UNMIL) who daily labor with us to keep the peace that we enjoy.

Fellow Liberians, Ladies and Gentlemen:

No one who has lived in or visited this country in the past fifteen years will deny the physical destruction and the moral decadence that the civil war has left in its wake here in Monrovia and in other cities, towns, and villages across the nation. We have all suffered. The individual sense of deprivation is immense. It is therefore understandable that our people will have high expectations and will demand aggressive solutions to the socioeconomic and societal difficulties that we face.

Our record and experience show clearly that we are a strong and resilient people, able to survive; able to rise from the ashes of civil strife and to start anew; able to forge a new beginning, forgiving if not forgetting the past. We are a good and friendly people, braced for hope even as we wipe away the tears of past suffering and despair. Our challenge, therefore, is to transform adversity into opportunity, to renew the promises upon which our nation was founded: freedom, equality, unity and individual progress. In the history of nations, each generation is summoned to define its nation's purpose and character. Now, it is our time to state clearly and unequivocally who we are, as Liberians—and where we plan to take this country in the next six years.

## POLITICAL RENEWAL

First, let me declare in our pursuit of political renewal, that the political campaign is over. It is time for us, regardless of our political affiliations and persuasions, to come together to heal and rebuild our nation. For my part, as President of the Republic of Liberia, my Government extends a hand of friendship and solidarity to the leadership and members of all political parties which participated in our recent presidential and legislative elections. I call upon those who have been long in the struggle—and those who recently earned their stripes—to play important roles in the rebuilding of our nation.

Committed to advance the spirit of inclusion, I assure all Liberians and our international partners and friends that our Government will recognize and support a strong democratic and loyal opposition in Liberia. This is important because we believe that our democratic culture and our nation are best served when the opposition is strong and actively engaged in the process of nation building.

Moreover, we call upon our colleagues of all political persuasions now in the Diaspora to return home and join us in meeting this exciting challenge of national renewal. We are aware that we have hundreds of doctors, engineers, and economists, as well as thousands of teachers, nurses, professors, and other Liberians who possess specialized skills currently living abroad. I re-echo my appeal to all of you to please come home!! Please make the sacrifice, for your country needs you and needs you now!!!

We make a similar appeal to the thousands of our citizens who continue to live in refugee camps throughout the sub-region and beyond. We recognize and sympathize with your plight and will explore with our development partners ways and means to facilitate your early return home as a national imperative for our renewal and development. To those who are still internally displaced, we pledge to work with our partners to get you back to your communities to enable you to start the process of rebuilding your lives.

We must have a new understanding. Your job, as citizens, is to work for your family and your country. Your country's only job is to work for you. This is the compact that I offer you today.

## A NEW ERA OF DEMOCRACY

My Fellow Liberians, Ladies and Gentlemen:

Ours has certainly not been an easy journey to where we are today. Indeed, the road has been tortuous and checkered. From the establishment of our National Polity 158 years ago through the period of integration which helped to shape our society several decades ago, to the long running political fight for the forging and fostering of a viable and cohesive society in the decades that followed, the struggle of many has centered on five core values: peace, liberty, equality, opportunity, and justice for all.

The tendencies of intolerance of each other's opinion rooted in parochial and selfish considerations—and greed—have driven us into our descent into recent tragedies and paralysis as a nation and people. These negative national tendencies have, in the past, bred ethnic suspicion and hatred, led to injustice, social and political exclusion. They have also weakened our capacity to peacefully co-exist as a people with diverse socio-cultural, economic, and political backgrounds and differences. Consequently, we have witnessed needless generalized conflicts that have profoundly affected the Liberian family, the foundation of our society.

And in the process of resolving the numerous contradictions that have underpinned this struggle, a high price has been paid by many Liberians of diverse backgrounds and social status. I know of this struggle because I have been a part of it. Without bitterness, anger, or vindictiveness, I recall the inhumanity of confinement, the terror of attempted rape, and the ostracism of exile. I also recall the goodness and the kindness of the many who defied orders and instruction to save my life, and give food to the hungry and to give water to the thirsty. I recall their humanity—and thank them.

And so, my Fellow Liberians, let us acknowledge and honor the sacrifices and contributions of all as we put the past behind us. Let us rejoice that our recent democratic exercise has been a redemptive act of faith and an expression of renewed confidence in ourselves. Let us be proud that we were able to ultimately rise above our intense political and other differences in a renewed determination as a people to foster dialogue instead of violence, promote unity rather than disharmony, and engender hope rather than disillusionment and despair.

Today, I urge all of us to commit ourselves to a new era of democracy in Liberia. In our new democracy, we will tolerate even if we disagree; we will co-exist even if we consider our neighbor unfriendly—and we will find common ground on the many vexing issues that face our nation. This is because our shared national values are more important than our individual interests. We must therefore abandon the temptation and inclination to court and engage in violence. Our recent history teaches us that violence diminishes our nation and ourselves, not just within our borders, but more importantly in our dealings with other nations and peoples.

My Administration therefore commits itself to the creation of a democracy in which the constitutional and civil liberties and rights of all our people will be advanced—and safeguarded. While ensuring the security of our nation and people, we will work tirelessly to ensure that the writ of democracy is expanded, not constricted in our land.

## Economic Renewal

In a similar quest for economic renewal, we start on the premise that we are a wealthy people. Our nation is blessed with an endowment rich in natural and human resources. Yet, our economy has collapsed due to the several civil conflicts and economic mismanagement by successive governments. The task of reconstructing our devastated economy is awesome, for which there will be no quick fix.

Yet, we have the potential to promote a healthy economy in which Liberians and international investors can prosper. We can create an investment climate that gives confidence to Liberian and foreign investors. We can promote those activities that add value in the exploitation of our natural resources. We can recognize and give support to our small farmers and our marketers who, through their own efforts over the years, have provided buoyancy and self-sufficiency in economic activity, even during the difficult years of conflict. We can revisit our land tenure system to promote more ownership and free holding for communities. We can expand ongoing programs of economic and social infrastructure rehabilitation.

This will call for the translation of our economic vision into economic goals that are consistent with our national endowment and regional and global dynamics. Included in this process is a formulation of the policy framework and identification of the sequential measures of structural change that need to be taken to achieve the overarching goals of sustainable growth and development. We will ensure that allocation of our own resources reflects these priorities. We will call upon our development partners to likewise recognize that although they have made significant investment to bring peace to our country, this peace can only be consolidated and sustained if we bring development to our people.

With this in mind, we are working with our partners to identify key objectives and deliverables in the first one hundred and fifty days of our Administration which coincides with the remaining budgetary period of the former government. We must meet our commitment to restore some measure of electricity to our capital city. We must put Liberians back to work again. And we must put our economic and financial house in order. Most of all, we must revive our mindset of courage, hard work, and a can-do spirit.

Our strategy is to achieve quick and visible progress that reaches a significant number of our people, to gain momentum, consolidate support, and establish the foundation for sustained economic development. This will encompass five major pillars: Security, Economic Revitalization, Basic Services, Infrastructure, and Good Governance.

In implementing the programs consistent with this strategy, we will ensure broad geographic representation and participation, placing emphasis on those areas that have received less in the distribution of economic benefits. In this regard, we thank the European Union for supporting activities in the power sector, in community development, as well as providing technical assistance in economic management. We thank the United States for supporting the restructuring and training of our security forces, for activities in community development, and for commencing the construction of the Barclayville Bridge in the neglected area of the Southeast.

We thank our sisterly countries of Nigeria and Ghana for providing training for our security forces. We thank the United Nations System for supporting community development, technical assistance for economic management, reintegration, and good governance. We thank them also for the strong peacekeeping effort, in conjunction with ECOWAS, which has enabled us to have an environment that led us to free and fair elections.

We thank the World Bank for support of activities in community development, infrastructure, and technical assistance in economic management. We thank the many foreign and domestic non-governmental organizations for their support for community development, and for peace and capacity building. As we look ahead, we plan to collaborate closely with both the international and national NGOs and the civil society community in order to formulate an appropriate strategy and approach for their engagement with our Government in order to maximize their contributions.

For the long term, more will be required from us and our partners. We will formulate a multi-year economic reconstruction plan tied to a Poverty Reduction Strategy Program that relieves our country from a staggering US$3.5 billion external debt and paves the way for acceleration in our national effort to make more progress in the achievement of the Millennium Development Goals. We will seek a strong partnership between the public and private sector, with direct foreign investment and Liberian entrepreneurship at the core. We envision a multi-year commitment in the order of US$1 billion for this purpose with details to be presented at a partnership meeting planned for May or June of this year.

As we seek to engage our youth in our enterprise of nation building, we must recognize the threat that HIV/AIDS poses to our human capital and to our growth and prosperity. With 12% of our population now affected by HIV/AIDS, my administration will tackle this national scourge by updating and reinvigorating our HIV/AIDS policy within our first 150 days. We will also reconstitute and empower, along with our development partners, the National Commission on HIV/AIDS.

## GOVERNANCE

We know that our desire for an environment for private sector–driven sustainable growth and development cannot be achieved without the political will and a civil service that is efficient, effective, and honest. The workforce in our ministries and agencies is seriously bloated. Moreover, many of the ministries and agencies lack clarity in mandate and have little or no linkages to our national priorities, policies, and goals. Our Administration will therefore embark on a process of rationalizing our agencies of government to make them lean, efficient, and responsive to public service delivery. This will require the creation of a meritocracy that places a premium on qualification, professionalism, and performance.

As a major component of our Civil Service Reform Agenda, we will review our public service wage system with the view to ensuring that those who work in our civil service are paid commensurate with their qualifications and performance—and that they are paid on time. It may take us some time to achieve this objective given our inheritance of a bloated and poorly paid civil service for which there are currently salary and benefit arrears totaling some US$20 million.

Our present unemployment situation is a national crisis. We must redeploy some of our current public service employees to areas where they can perform successfully. We will start the process to train and retrain others who lack requisite professional skills. We will empower them through our proposed alternative employment initiatives. We will also provide additional support through our proposed micro-loan program.

## BONDING

My Fellow Liberians, Ladies and Gentlemen:

Across this country, from Cape Mount in the West to Cape Palmas in the East, from Mount Nimba in the North to Cape Monsterrado in the South, from Mount Wologizi in Northcentral to Mount Gedeh in the Southeast, our citizens at this very moment are listening to my voice by radio—and some are watching by television. I want to speak directly to you. As you know, in our various communities and towns, our children have a way of greeting their fathers when they come home after a long, tiring day of trying to find the means to feed the family that night and send the children to school the next day. They say, "Papa na come."

Well, too many times, for too many families, Papa comes home with nothing, having failed to find a job or to get the help to feed the hungry children. Imagine then the disappointment and the hurt in the mother and children; the frustration and the loss of self-confidence in the father. Through the message of this story, I want you to know that I understand what you, our ordinary citizens, go through each day to make ends meet for yourselves and for your families.

Times were hard before. Times are even harder today. But I make this pledge to you: Under my Administration, we will work to change that situation. We will work to ensure that when our children say "Papa na come," Papa will come home joyfully with something, no matter how meager, to sustain his family. In other words, we will create the jobs for our mothers and fathers to be gainfully employed. We will create the social and economic opportunities that will restore our people's dignity and self-worth.

We will make the children smile again; the thousands of children who could not present their voting cards, but repeatedly told me whenever I met and shook their hands that they voted for me. Indeed, they voted with their hearts. To those children and to all other Liberian children across this nation, I say to you, I love you very, very much. I shall work, beginning today, to give you hope and a better future.

Now, I would like to speak in particular to our youth. You can believe my word that my Administration will do its utmost to respond to your needs. We will build your capacity and empower you to enable you to meaningfully participate in the reconstruction of our country. We shall actively pursue the Kakata Declaration resulting from the National Youth Conference held in 2005 and the implementation of a National Youth Policy and Program.

## CORRUPTION

Fellow Liberians, we know that if we are to achieve our economic and income distribution goals, we must take on forcibly and effectively the debilitating cancer of corruption. Corruption erodes faith in government because of the mismanagement and misapplication of public resources. It weakens accountability, transparency, and justice. Corruption shortchanges and undermines key decision and policy making processes. It stifles private investments which create jobs and assures support from our partners. Corruption is a national cancer that creates hostility, distrust, and anger.

Throughout the campaign, I assured our people that, if elected, we would wage war against corruption regardless of where it exists, or by whom it is practiced. Today, I renew this pledge. Corruption, under my Administration, will be the major public enemy. We will confront it. We will fight it. Any member of my Administration who sees this affirmation as mere posturing, or yet another attempt by yet another Liberian leader to play to the gallery on this grave issue should think twice. Anyone who desires to challenge us in this regard will do so at his or her personal disadvantage.

In this respect, I will lead by example. I will expect and demand that everyone serving in my Administration leads by example. The first testament of how my Administration will tackle public service corruption will be that everyone appointed to high positions of public trust such as in the Cabinet and heads of public corporations will be required to declare their assets, not as part of a confirmation requirement, but as a matter of policy.

I will be the first to comply by declaring my assets. My Administration will also accord high priority to the formulation and passage into law of a National Code of Conduct, to which all public servants will be subjected.

My Fellow Liberians, Ladies and Gentlemen:

If we are to achieve our development and anti-corruption goals, we must welcome and embrace the Governance and Economic Management Program (GEMAP) which the National Transitional Government of Liberia, working with our international partners, has formulated to deal with the serious economic and financial management deficiencies in our country.

We accept and will enforce the terms of GEMAP, recognizing the important assistance which it is expected to provide during the early years of our Government. More importantly, we will ensure competence and integrity in the management of our own resources and insist on an integrated capacity-building initiative so as to render GEMAP non-applicable in a reasonable period of time.

## FOREIGN POLICY

My Fellow Liberians:

Our nation's foreign policy has historically been rooted in our core values as a nation and people in the practices of good neighborliness, non-interference in the affairs of other nations and peoples, peaceful co-existence, regional cooperation and integration, and international bilateral and multilateral partnership. These core values will continue to guide the conduct of our foreign policy under my Administration.

Our foreign policy will take due cognizance of the sacrifices and contributions that have been made to restore peace, security, and stability to our country. We will therefore work to be a responsible member of sub-regional, regional, and international organizations, including the Mano River Union, Economic Community of West African States (ECOWAS), African Union, and the United Nations. We will do all that we can to honor our obligations, past and current, and enforce all international treaties to which our country has subscribed.

To our sister Republics West, East, and North of our borders, we make this pledge: under my Administration, no inch of Liberian soil will be used to conspire to perpetrate aggression against your countries. In making this commitment, we will work for a new regional security that is based upon economic partnership aimed at enhancing the prospects for regional cooperation and integration. In this regard, we propose to tap into the successful Southeast Asian experiences regarding the promotion of regional integration with a heavy private sector component.

I also want to speak specifically to the countries which, from the onset of our civil conflict, have been in the vanguard of peace-making in Liberia. Some have made material and financial contributions. Some have provided moral support. Others have contributed troops that paid the supreme price for peace in our country. To our war dead, our brother soldiers from West Africa and other regions—as well as to our own, we remember and honor you today. This occasion is owed to your fortitude and to your sacrifices. To every other nation and partner, we thank you for standing by us.

## RECONCILIATION

Today, as we usher in a new era of responsibility, accountability, and transparency, we must strive to reawaken our people's faith in their Government. We must also recognize the urgency and imperative of meeting the challenges of post-conflict reconstruction. Yet, no single issue or factor will define our success or failure in this endeavor more than our willingness and ability to come together as a nation and people. Consequently, no task will be more urgent and more compelling, no cause will require my personal attention and engagement more than national reconciliation.

As in the case of the overall challenge of economic reconstruction, there will be no quick fix to national reconciliation and healing. But we can neither flinch from the challenge, nor be overwhelmed by its complexities. After all, some of the underlying factors of our current problems are as deep and old as the history of our country. So, we must begin today to reconcile and heal our nation with deliberate and purposeful commitment, recognizing that we are first and last Liberians— and that our nation's strength, progress, and development are directly impacted by our unity, peace, security, and stability as a people.

Therefore, I today pledge my personal involvement in the work of reconciling and healing our country. The Truth and Reconciliation Commission has an important role to play in this regard and my Administration will support and strengthen the Commission to enable it to carry out its mandate effectively.

My Fellow Citizens:

Let me assure you that my Presidency shall remain committed to serve all Liberians without fear or favor. I am President for all of the people of this country. I therefore want to assure all of our people that neither I nor any person serving in my Administration will pursue any vendetta. There will be no vindictiveness. There will be no policies of political, social, and economic exclusion. We will be an inclusive and tolerant Government, ever sensitive to the anxieties, fears, hopes, and aspirations of all of our people irrespective of ethnic, political, religious affiliations, and social status. Let us be clear, however, that we will insist on specified standards of law abiding behavior in the exercise of this tolerance.

My Fellow Liberians, Ladies and Gentlemen:

By their votes, the Liberian people have sent a clear message! They want peace; they want to move on with their lives. My charge as President is to work to assure the wishes of our people. We will therefore encourage our citizens to utilize our system of due process for settling differences whether those differences are within or between ethnic groups, or whether they are within or between religious groups. However, we will forcefully, swiftly, and decisively respond to any acts of lawlessness, threats to our hard-earned peace, or destabilizing actions that could return us to conflict.

As we today savor the new dawn of hope and expectation, I pledge to bring the Government closer to the people. The days of the imperial Presidency, of an intrusive leadership, and of a domineering and threatening Chief Executive are over in Liberia. This was my campaign promise which I intend to keep. Yet, my Government will be unflinching and bold in influencing and defending those measures that ensure that our national goals are achieved.

In pursuing this policy, our Constitution will remain our source of strength. Its edifying phrase, WE, THE PEOPLE OF THE RE-PUBLIC OF LIBERIA, and its equally ennobling proclamation that ALL POWER IS INHERENT IN THE PEOPLE, will be given concrete meaning and expression in all of our national life and conduct. The Executive Mansion and Monrovia will no longer be the only centers of power and sources of development policy making. The people and their interests, as defined by them, will be at the very heart of our new dispensation of decentralization and the devolution of power.

And now I would like to talk to the women, the women of Liberia, the women of Africa—and the women of the world. Until a few decades ago, Liberian women endured the injustice of being treated as second-class citizens. During the years of our civil war, they bore the brunt of inhumanity and terror. They were conscripted into war, gang raped at will, forced into domestic slavery. Yet, it is the women, notably those who established themselves as the Mano River Women's Network for Peace who labored and advocated for peace throughout our region.

It is therefore not surprising that during the period of our elections, Liberian women were galvanized—and demonstrated unmatched passion, enthusiasm, and support for my candidacy. They stood with me; they defended me; they prayed for me. The same can be said for the women throughout Africa. I want to here and now, gratefully acknowledge the powerful voice of women of all walks of life whose votes significantly contributed to my victory.

My Administration shall thus endeavor to give Liberian women prominence in all affairs of our country. My Administration shall empower Liberian women in all areas of our national life. We will support and increase the writ of laws that restore their dignities and deal drastically with crimes that dehumanize them. We will enforce without fear or favor the law against rape recently passed by the National Transitional Legislature. We shall encourage families to educate all children, particularly the girl child. We shall also try to provide economic programs that enable Liberian women to assume their proper place in our economic revitalization process.

My Fellow Liberians:

We are moving forward. Our best days are coming. The future belongs to us because we have taken charge of it. We have the resources. We have the resourcefulness. Now, we have the right Government. And we have good friends who want to work with us. Our people are already building our roads, cleaning up our environment, creating jobs, rebuilding schools, bringing back water and electricity.

My Government will ensure that the creativity and industry of Liberians is unleashed in this incredible moment of history. We are making our beloved Liberia home once again. We are a good people; we are a kind people. We are a forgiving people—and a God-fearing people. So, let us begin anew, moving forward into a future that is filled with hope and promise! ". . . In Union Strong, Success is Sure! We cannot fail . . ." God bless us all—and save the Republic.

I Thank You!

# ACKNOWLEDGMENTS

THANKS TO:

My sister, Jennie Bernard, who helped refresh my memory of the countless childhood experiences we shared. You've always given so freely without regard for what you'd get in return. Please know that my story, in so many ways, is a joyous one because of you.

Clavenda Bright Parker, a true friend and confidant, for helping me convey intimate details of some of the more challenging times in my life. Thanks for encouraging and pushing me to share the whole truth about my journey.

My son Rob, for his continuous involvement in and commitment to this project, staying the course with deadlines and managing all that I couldn't. I obviously taught you well.

Finally, Molly Cashin, thanks for your tireless work editing and reviewing this manuscript. The insight and suggestions you provided are vital to this book's success.

# BIBLIOGRAPHY

## NEWSPAPERS

*Australian,* various dates.
*Christian Science Monitor,* various dates.
*Financial Times,* various dates.
*Guardian,* various dates.
*New Internationalist,* January 1984.
*New York Times,* various dates.
*Philadelphia Inquirer,* various dates.
*Sunday Times* (London), various dates.
*Washington Post,* various dates.

## WEB SITES AND DATABASES

"The African-American Mosaic: A Library of Congress Resource Guide for the Study of Black History & Culture," www.loc.gov/exhibits/african/afam002.html.

"American Colonization Society," www.globalsecurity.org/military/library/report/1985/liberia_1_americancolsoc.htm.

Data & Information Services Center, Online Data Archive. "Roll of Emigrants to Liberia, 1820–1843 and Liberian Census Data, 1843," Tom W. Shick, Principal Investigator; www.disc.wisc.edu/liberia.

"The Liberian People," http://pages.prodigy.net/jkess3/People.html.

Moran, Mary H. "Culture of Liberia," www.everyculture.com/Ja-Ma/Liberia.html.

Shick, Tom W. "Emigrants to Liberia, 1820–1843," www.disc.wisc.edu/liberia/pdfs/alphalist.html.

van der Kraaij, Fred P. M. "Liberia: Past and Present of Africa's Oldest Republic," www.liberiapastandpresent.org.

## BOOKS AND JOURNALS

Banks, Russell. *The Darling: A Novel.* New York: HarperPerennial, 2005.

Boley, G. E. Saigbe. *Liberia: The Rise and Fall of the First Republic.* New York: St. Martin's Press, 1983.

Kocher, Kurt Lee. "A Duty to America and Africa: A History of the Independent African Colonization Movement in Pennsylvania." *Pennsylvania History* 51 (April 1984).

Liebenow, J. Gus. *Liberia: The Evolution of Privilege.* Ithaca, N.Y.: Cornell University Press, 1969.

———. *Liberia: The Quest for Democracy.* Bloomington: Indiana University Press, 1987.

Massachusetts Colonization Society. *American Colonization Society and the Colony at Liberia.* Ithaca, N.Y.: Cornell University Library, 1832.

Pham, John-Peter. *Liberia: Portrait of a Failed State.* New York: Reed Press, 2004.

Shick, Tom W. *Behold the Promised Land: A History of Afro-American Settler Society in Nineteenth-Century Liberia.* Baltimore: Johns Hopkins University Press, 1980.

Tuck, Christopher. "'Every Car or Moving Object Gone': The ECOMOG Intervention in Liberia." *African Studies Quarterly: The Online Journal for African Studies,* 4(1): 1; http://web.africa.ufl.edu/asq/v4/v4i1a1.htm.

Wiley, Bell I., ed. *Slaves No More: Letters from Liberia, 1833–1869.* Lexington: University Press of Kentucky, 1980.

# INDEX